MANAGEMENT SERIES

MANAGING THE ENVIRONMENT
IN
CRITICAL CARE NURSING

The First Century 1890-1990

MANAGEMENT SERIES

MANAGING THE ENVIRONMENT
IN
CRITICAL CARE NURSING

EDITED BY

Joan Gygax Spicer, RN, MSN, MBA, CNAA

Assistant Director
Department of Nursing
The Medical Center at the University of California, San Francisco
San Francisco, California

MaryAnne Robinson, RN, MS

Case Manager
Acute Care
The Medical Center at the University of California, Irvine
Orange, California

WILLIAMS & WILKINS
Baltimore • Hong Kong • London • Sydney

Editor: Susan M. Glover
Associate Editor: Marjorie Kidd Keating
Copy Editor: Shelley Potler
Designer: Saturn Graphics
Illustration Planner: Wayne Hubbel
Production Coordinator: Anne Stewart Seitz

Copyright © 1990
Williams & Wilkins
428 East Preston Street
Baltimore, Maryland 21202, USA

All rights reserved. This book is protected by copyright. No part of this book may be reproduced in any form or by any means, including photocopying, or utilized by any information storage and retrieval system without written permission from the copyright owner.

Printed in the United States of America

Library of Congress Cataloging in Publication Data

Managing the environment in critical care nursing/edited by Joan
　Gygax Spicer, MaryAnne Robinson.
　　　p. cm.—(AACN management series)
　Includes bibliographical references.
　ISBN 0-683-07891-7
　1. Intensive care. I. Spicer, Joan Gygax. II. Robinson,
MaryAnne. III. Series.
　　[DNLM: 1. Critical Care—nursing. 2. Intensive Care Unit—
organization & administration. 3. Nursing Service, Hospital—
organization & administration. WY 154 M2665]
RT120.I5M362　1990
610.73′61—dc20
DNLM/DLC
for Library of Congress　　　　　　　　　　　　　　90-12040
　　　　　　　　　　　　　　　　　　　　　　　　　　　CIP

90 91 92 93 94
1 2 3 4 5 6 7 8 9 10

We dedicate this book to our parents:
the late Roy G. Gygax
and
Lucille Baumgartner Gygax
Waukesha, Wisconsin

the late Pablo Caigoy
and
Hermenigilda Caigoy
El Paso, Texas

who inspired the directions of
our personal and professional journeys

Foreword

Managing the Environment in Critical Care Nursing is the American Association of Critical-Care Nurses (AACN) second book in a series of management books for critical care nurses. This series is based on the AACN document, "Role Expectations for the Critical Care Manager." This second book in the series was developed to expand on the area of the manger's role and functions specific to management of the environment. Management in the critical care environment is important if there is to be effective integration of the critically ill patient, the role of the critical care nurse, and the critical care environment so that competent nursing practice and optimal patient outcomes occur.

As described in the "AACN Scope of Critical Care Nursing Practice," the critical care environment can be viewed from three perspectives: the global view, the setting, and the immediate environment. The broadest perspective of the environment encompasses a global view of factors in the legal, regulatory, social, economic, and political areas that have potential implications for critical care nursing. The institution or setting within which critically ill patients receive care represents another aspect of the critical care environment. Mangement and administrative structures within the institution must ensure effective care delivery systems for different populations of critically ill patients. The immediate environment must constantly support the interaction between the critical care nurse and the critically ill patient in order to effect desired patient outcomes.

Joan Spicer and MaryAnne Robinson have addressed perspectives of the critical care environment. Within AACN's conceptual model of the environment they have further differentiated that environment into layers or spheres providing a more detailed framework in which existing knowledge and research can be organized. They have identified strategic and administrative programs, support systems, and physical and human resources spheres. The specific content of the book is organized within these areas and includes the external environment, controlling quality, infection control, safety, clinical research, management of space, materiel, and equipment. Other topics include roles and relationships and ancillary and support services. Discussion of the patient and family in relationship to the environment is also included.

The authors contributing to this book come from a variety of settings and backgrounds. They share their knowledge, expertise, and personal insights on management of the environment. The strength of this book is that it originates from the professionals in the field. These authors are involved every day at various levels with management of the environment that has a direct or indirect impact on nurses and patients. The editors and contributing authors are to be commended for providing vital information necessary to assist critical care nurses in successful management in the next decade.

Suzanne K. White, RN, MN
President, American Association of
   Critical-Care Nurses
Assistant Director of Nursing/
   Director of Cardiovascular Nursing
Emory University Hospital
Assistant Professor
Nell-Hodgson Woodruff School of Nursing
Atlanta, Georgia

Preface

Managing the Environment in Critical Care Nursing provides a concise reference on current practices in effective environmental management. Each chapter contains a review of the related theory and practical applications for use in critical care work settings.

In Chapter 1, a conceptual model of the critical care work environment is described and the nursing research related to management of the environment is included in the discussion. Although the concept of environmental management is not new, application to the hospital environment and, therefore, to nursing is relatively new. The environment is differentiated into five spheres, and these are used as the framework to organize the presentation of the following chapters.

Chapters 2 and 3 describe the strategic sphere that is the outer limit for the conceptual model of the environment. The strategic sphere represents the environment in which health care, nursing, and organizations must coexist. The strategic sphere encompasses relationships between external environments and the critical care unit. Within the strategic sphere, processes and activities occur that are future directed and that can introduce the objectives that will bring about the desired future state.

Chapters 4, 5, and 6 discuss the administrative program sphere. The administrative program sphere includes those structures that establish the policies under which the continuing effectiveness of critical care nursing services are evaluated. Administrative program structures provide a driving force in environmental management. The administrative program structures include quality control, quality assurance and risk management, infection control, and clinical nursing research.

Chapters 7, 8, and 9 present the support systems sphere including ancillary and support department functions, integrated materiel support systems, and computer technology.

Chapters 10 and 11 discuss the physical sphere. This is the most concrete part of the environment and encompasses the controllable dimensions that are responsive to physical and spatial manipulation. Discussions cover space, equipment, and safety.

Chapters 12, 13, and 14 present the human resources sphere, which is the largest and most complex component of the environment and which transcends all the other spheres in the environment. The human resources sphere has an impact on every aspect of the environment. Roles and relationships are discussed. The critical care patient and the family are identified within this sphere as an integral part of the environment.

Effective management of the environment extends the critical care nurse's effectiveness and, therefore, has an impact on patient care outcome. If the knowledge gained by the reader through the use of this book is appropriately applied in the critical care work environment, patient care outcome, staff nurse job satisfaction, and efficiency and efficacy of delivery of critical care nursing services will all be enhanced.

Joan Gygax Spicer, RN, MSN, MBA, CNAA
MaryAnne Robinson, RN, MS

Acknowledgments

We thank our families for supporting us in this effort. Thank you Bruce Spicer and Nugent, Charles, Anne Marye, and Douglas Robinson.

We are grateful to the following people for their professional support in preparing this text:

- The contributing authors for sharing their knowledge, expertise, and insights;
- Margo Sorenson, BA, University of California, Los Angeles, English Department Chair, Harbor Day School, Corona del Mar, California, for providing editorial review of each chapter;
- Ann Bastovan, Administrative Assistant, Stanford University, for word processing support; and
- Elizabeth Yznaga Holliday, MSN, CNM, James Holliday & Associates, Graphics Division, for creating the graphics for the text.

Contributors

Judith Ann Blaufass, RN, MS
Director of Critical Care Nursing
LDS Hospital
Salt Lake City, Utah

Gladys M. Campbell, RN, MSN
Head Nurse, Cardiac Surgical Intensive Care Unit
National Institutes of Health Clinical Center
Bethesda, Maryland

Marianne Chulay, RN, DNSc, CCRN
Clinical Nurse Specialist, Cardiac Surgical Intensive Care Unit
National Institutes of Health Clinical Center
Bethesda, Maryland

Pamela F. Cipriano, RN, MN
Assistant Professor
Medical University of South Carolina
College of Nursing
Charleston, South Carolina

Lawrence A. Davidson
Logistics Manager and Quality Assurance Coordinator
Bendix Field Engineering Corporation
Suisun City, California

Kathleen E. Ellstrom, RN, MS, CCRN
Clinical Nurse Specialist/Nurse Manager
Medical Intensive Care Unit
University of California, Irvine Medical Center
Adjunct Faculty; California State University
Long Beach, California

Grandee R. Hardy, RN, MSN, CNA
Clinical Director of Cardiovascular Nursing Service
Administrator, MUSC Heart Center
Medical University of South Carolina
Charleston, South Carolina

George Hickey
Certified Clinical Engineer
George Hickey and Associates
Santa Monica, California

Lynne Kreutzer-Baraglia, RN, MS
Assistant Professor
West Suburban College of Nursing
Oak Park, Illinois

Karen Logsdon
Senior Vice President for Professional Services
California Association of Hospitals and Health Systems
Sacramento, California

Margaret M. Macari-Hinson, RN, MSN
Clinical Nurse Specialist
Intensive Cardiac Care and Cardiac Arrhythmia Monitoring Unit
University of California, Irvine Medical Center
Orange, California

Diane E. Nitta, RN, MA
Associate Administrator Patient Care Services/Director of Nursing Service
Valley Medical Center of Fresno
Fresno, California

Susan G. Osguthorpe, RN, MS, CCRN, CN
Assistant Chief Nursing Service/Surgery
Veterans Administration Medical Center
Salt Lake City, Utah

Beverly A. Post, RN, MS, CIC
Clinical Risk and Infection Control Manager
Louis A. Weiss Memorial Hospital
Chicago, Illinois

Claire E. Sommargren, RN, CCRN
Staff Nurse, Critical Care Unit
Dominican Santa Cruz Hospital
Santa Cruz, California

Suzanne K. White, RN, MN
President, American Association of Critical-Care Nurses
Assistant Director of Nursing/Director of Cardiovascular Nursing
Emory University Hospital
Assistant Professor
Nell-Hodgson Woodruff School of Nursing
Atlanta, Georgia

Contents

Foreword vii
Preface ix
Acknowledgments xi
Contributors xiii

PART I: INTRODUCTION

1 Conceptual Model of the Critical Care Work Environment 2
   Joan Gygax Spicer, MSN, MBA, RN, CNAA
   MaryAnne Robinson, MS, RN

PART II: STRATEGIC SPHERE

2 External Environment: Forces and Trends Reshaping Critical Care Nursing 12
   Karen Logsdon
3 Strategic Planning: Designing a Future 13
   Joan Gygax Spicer, MSN, MBA, RN, CNAA
   MaryAnne Robinson, MS, RN

PART III: ADMINISTRATIVE SPHERE

4 Controlling Quality in the Critical Care Environment 30
   MaryAnne Robinson, MS, RN
   Joan Gygax Spicer, MSN, MBA, RN, CNAA
5 Infection Control 42
   Beverly A. Post, MS, RN, CIC
   Lynn Kreutzer-Baraglia, MS, RN
6 Establishing a Clinical Nursing Research Program 52
   Gladys M. Campbell, MSN, RN
   Marianne Chulay, DNSc, RN, CCRN

PART IV: SUPPORT SYSTEMS SPHERE

7 Ancillary and Support Services 62
   Susan G. Osguthorpe, MS, RN, CCRN, CN
8 Integrating Materiel Support Systems 82
   Lawrence A. Davidson
   Joan Gygax Spicer, MSN, MBA, RN, CNAA
9 Computer Technology 93
   Judith Ann Blaufuss, MS, RN

PART V: PHYSICAL SPHERE

10 Physical Plant Design and Equipment Procurement 106
    Diane E. Nitta, MA, RN
    George Hickey
11 Occupational Safety 120
    Claire E. Sommargren, RN, CCRN

PART VI: HUMAN RESOURCES SPHERE

12 Roles and Relationships 140
      Pamela F. Cipriano, MN, RN
      Grandee R. Hardy, MSN, RN, CCRN, CNA

155 Customer Relations: A Supportive Environment for the Family 156
      Joan Gygax Spicer, MSN, MBA, RN, CNAA
      MaryAnne Robinson, MS, RN

14 The Critical Care Patient: Characteristics and External Environmental Influences 162
      Kathleen E. Ellstrom, MS, RN, CCRN
      Margaret Macari-Hinson, MSN, RN

Appendices
    A 173
    B 174
    C 176
    D 178

Index 181

Part I

INTRODUCTION

Chapter 1
Conceptual Model of the Critical Care Work Environment

JOAN GYGAX SPICER, MARYANNE ROBINSON

In optimally managed critical care environments, the patient receives efficient, necessary, and appropriate nursing services; the members of the nursing staff experience increased job satisfaction; and the organization achieves success in the marketplace through patient, physician, and third party payor satisfaction, in addition to increased productivity. The purpose of this chapter is to define the critical care environment, to diagram a conceptual model of the critical care environment, and to review pertinent nursing literature and nursing research. The intent of this chapter is to provide nurse managers with a basis on which to analyze the effectiveness of management decisions that have an impact on the critical care environment.

Environment is defined as being the point where interaction between groups (people-people) and interface between material resources and human resources (things-people) occurs. A conceptual model provides a diagram of the elements composing the critical care environment. It represents a dynamic flowing process of the actual or potential interactions and interfaces of the elements within the environment. Nurse managers' decisions have an impact on when and how the elements interrelate.

Nurse managers are keenly aware of the role they play in balancing the complexities within the environment; therefore, it is important to diagram conceptually the critical care environment. This provides a framework within which nurse managers can describe and discuss the interrelatedness of these elements in the environment. These discussions can create shared understanding of the factors involved in managing the environment.

As nurse managers gain insights into the interrelatedness of the elements in the environment, they can analyze their decisions, postulating why the planned impact was or was not accomplished and predicting to the outcomes of their future decisions. As hunches and hypotheses are formulated, the model can then be grounded in theory and scientific knowledge. Research can validate correlations, causes, and effects of the interrelatedness of the elements in the environment.

Stevens (1) was one of the first authors to put forth a conceptual model of the role of the head nurse and the impact of the head nurse's actions on the elements in the environment. Stevens identified five elements in the environment: staff, systems, self [head nurse], patients, and administration, and discussed the interrelatedness of these elements (1). Ganong and Ganong (2) depicted the work world of the head nurse and identified in their diagram linkages with environmental factors of business, economic, professional, governmental, community, political, and social factors as well as the components of the medical staff, hospital administration, employees, other departments, patient families, and the culture. Both of these early models identified the rudiments of the elements within the environment. The authors postulated that the effectiveness of head nurse role hinged on the impact that the head nurse's actions had on the interrelatedness of the environmental elements as identified by the authors, respectively, and that the environmental elements had an impact on the head nurse role as well.

Other nurse-authors have discussed conceptual models of the work environment. Simms et al. (3), in the Health Organization-Environment Model, described the work environment of the nurse as having an internal and external layer. The internal layer includes three parts: structure components; process components; and outcome components. Structure components include such things as a philosophy statement, policies, nursing assignment patterns, and job descriptions. Examples of process components are nursing orders, patient care criteria and standards, and clinical judgments. Such factors as trust, support, and degree of risk-taking are included as the outcome components (3). These authors affirm the importance of developing a conceptual model in order to articulate management behaviors better and to improve and validate nursing practice.

Kinney et al. were some of the first nurses to provide a model to describe the critical care environment and its relationship to the critical care nurse, the critically ill patient, and the nurse-patient interaction. This model was adopted by the American Association of Critical-Care Nurses (AACN) Board of Directors, November 1986 in its position statement on, "Scope of Critical Care Nursing Practice" (4). The AACN model is composed of three levels: (*a*) factors influencing behaviors/practices at the patient care level, (*b*) factors influenced by institutional behavior and values, and (*c*) factors influenced by global external behaviors and policy (4). The AACN model (Fig. 1.1) is an important contribution to the description of the critical care environment in relationship to the care of the critically ill patient.

NURSING LITERATURE AND RESEARCH ON MANAGEMENT OF THE ENVIRONMENT

The earliest hypotheses correlating the importance of the environment to patient care were put forward by Florence Nightingale who, through data collection, showed connections among diseases, sanitation, organization designs, and the management of health ser-

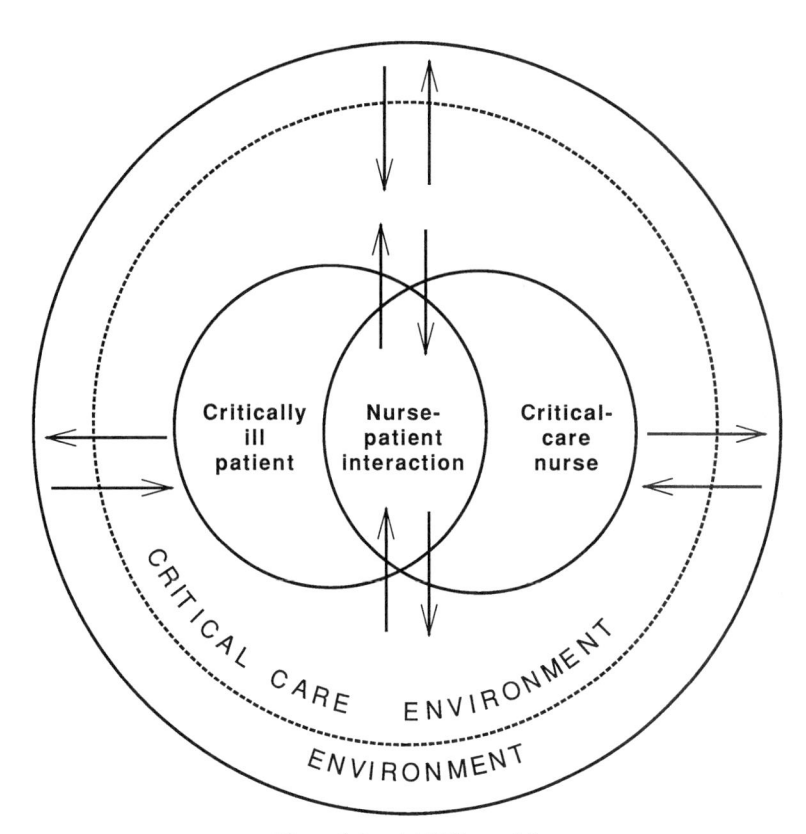

Figure 1.1. AACN's model.

vices for specific patient populations (5). Still, the need for research on the interrelatedness of the elements of the environment and patient care outcomes exists after all these years.

Research on the interrelatedness of elements in the critical care environment is limited. The relationships initially described in the literature more often than not are either hunches or are based on intuition. There have been only a few landmark research studies published in reference to relationships of elements in the critical care environment to patient care outcomes.

William's (6) review of the literature related to environment and patient care identified research done by several disciplines focusing on the interrelatedness between human behavior and broad aspects of the environment. Research related to physical environment and patient care outcomes were nontheoretically based because specific data base studies were few. Williams concluded that research relating environment to patient care outcomes has not been emphasized. Furthermore, research in this area by the nursing discipline has been very limited.

Mitchell et al. (7) measured and categorized various environmental factors such as organizational structures and professional sociological processes and attempted to correlate these with organizational attributes and patient care outcomes. This study supported the plausibility of manipulating aspects of the critical care environment in order to influence the quality of care. This study concluded that positive organizational and clinical outcomes coexist with valued aspects of the environment (7).

Knaus et al. (8) supported the hypothesis that the degree of coordinated services in an intensive care unit significantly influenced its effectiveness. Outcomes for patients in critical care units were more related to the interaction and coordination of each hospital's staff members than to the administrative structure, the amount of specialized treatment used, or to the hospital's teaching status (8).

The result of the American Academy of Nursing research on "Magnet Hospitals" was published in 1983. This study (9) linked aspects of internal organization processes to positive characteristics of environmental management. It was a descriptive study of the environments that attract and retain professional nurses. It was not designed to study the impact of environmental structures on patient care outcomes, much less to do so concerning critical care. It is a significant study, however, and bears inclusion in discussions related to managing the critical care environment, if it is assumed that an organization's ability to recruit and retain nurses affects the ability to deliver consistent nursing services for groups of patients.

The National Commission on Nursing was a group of leaders from nursing, medicine, hospital administration, education, government, business, and boards of trustees of health care organizations. The Commission was charged with examining problems and issues affecting nursing and with recommending actions for the future. Among the recommendations from this commission, at least five reflect the essence of management of the environment and its applicability to the critical care nurse manager's role:

- Nurses have an obligation to establish a suitable environment for nursing practice.
- There must be participation in a collaborative practice relationship with the physician to assure quality patient care.
- There must be recognition of nursing practice as its own clinical practice discipline having authority over its management processes.
- Qualifications of nurse managers should include the ability to promote, develop, and maintain an environment conducive to managing nursing resources and quality of care.
- Nurses must be actively involved in establishing standards and evaluating the quality of support services (10).

CONCEPTUAL MODEL OF THE CRITICAL CARE ENVIRONMENT

Building on AACN's Model of the Critical Care Environment, Spicer and Robinson propose to differentiate the environment further (Fig. 1.2). In this model, the environment is differentiated into five layers or spheres: (*a*) strategic, (*b*) administrative programs, (*c*) support systems, (*d*) physical, and (*e*) human resources. An interrelatedness of the different layers or spheres is created by the interchanges or conditions

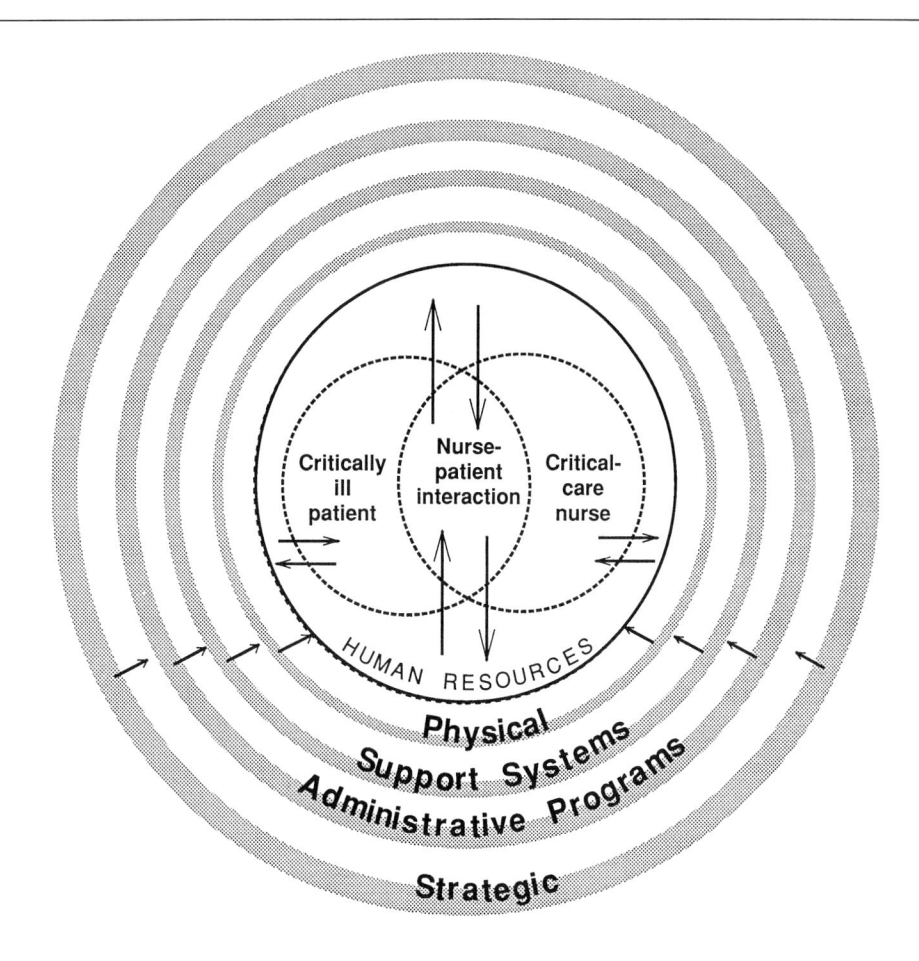

Environmental Management

Figure 1.2. Differentiating spheres of the environment.

within each sphere. Specific factors within each sphere are controllable through management interventions.

Strategic Sphere

The strategic sphere encompasses relationships between external environments and the critical care work environment. The strategic sphere is the outer limit of the conceptual model, which could be infinity. It consists of the economic, social, legal, technical, and professional conditions or climates in which nursing exists. Strategic planning is the process of placing the macroscopic view of the external environment in a perspective relative to the patient, nurse, and practice environment.

Before the 1980s, and the Tax Equity and Fiscal Responsibility Act of 1982, health care was based on a fee-for-service and retrospective payment model existing in an inelastic market; no matter what the price, the demand for the health care services was unchanged. As the fee-for-service and retrospective payment model has been challenged and capped, the prospective payment model came into being. Thus, health care organizations were forced to compete in an quasielastic marketplace based on price, meeting health care needs, and providing technology. It is critical that organizations functioning in a competitive marketplace know there are documented cases of what happens when administrators and managers do not interpret the environment and design a future based on that information, i.e., success in the marketplace is not achieved.

The 1983 Nursing Commission Report validates the importance of internal linkage to the external environment in order to meet the demands of today's health care. Nursing services that were required 20 years ago cannot meet the demands of health care needs today. The Commission (10) summarized the need for nursing to link up its internal operations to its external environment because the demands of health care services require specially skilled and knowledgeable nurses to provide unique and specialized care. The linkage of the critical care environment to its external environment occurs in the strategic sphere where the external processes indirectly have an impact on conditions of practice and define limits within the work environment. It is also within this sphere that nurses strategize how to make nursing care a competitive advantage in the marketplace for their organization.

Administrative Program Sphere

The administrative program sphere includes those structures that establish the policies under which the continuing effectiveness of the total operations are evaluated. Administrative program structures provide a driving force in management of the environment. The administrative program structures are outcome oriented and include factors such as quality assurance and risk management, infection control, and clinical nursing research. These programs measure achievement of patient care outcomes and estimate the impact the elements of the environment had on the process of achieving these outcomes. Activities of the program may include recommending how to manipulate the elements in the environment in order to benefit the patient further.

In the past decade, quality assurance and risk management programs have become a primary focus in the administrative sphere. Technology advances, an aging population, catastrophic diseases, and the like have increased liability for the hospital and the caregiver. Third party payors and the federal government are taking a hard look at utilization and appropriateness of services, particularly in areas such as critical care units where resources are easily and quickly absorbed.

Clinical research in the critical care environment is in its infant stage. Although much is written on different aspects of critical care procedures and nursing practice, utilizing research data base applications is weak. Researchers (6–8) concur that more data base research is needed to validate findings related to patient care outcomes and environmental factors, such as organizational attributes.

Nursing management practice research is in direct partnership with clinical nursing practice research. Henry et al. (11) defined nursing administration research as that research "concerned with establishing the costs of nursing care, with examining the relationships between nursing services and quality patient care, and with viewing problems of nursing service delivery within the broader context of policy analysis and delivery of health care services." The most important research question identified was the validation of the cost-effectiveness of clinical nursing practice and its benefits in promoting consumer satisfaction, patient care outcomes, and cost of quality (11).

Support Systems Sphere

The support systems sphere is the interaction and interfacing among the critically ill patient, the human resources, the support and ancillary functions, and the materiel resources. This interface also occurs between human resources and high technology and automation.

Ancillary and support functions are crucial to the critical care environment because of the critically ill patient's needs. The Society of Critical Care Medicine (SCCM) Task Force (12) outlined the necessity of having pathology, pharmacy, radiology, and respiratory therapy support on a 24-hour basis in order to provide efficient and necessary critical care services to the patient. The National Commission identified that, "The presence and quality of supporting services to the patient care unit is a major determinant in the effectiveness of the delivery system and the satisfaction of the professionals working in the system" (10). This was substantiated in the Magnet Hospital study (9) where nurses identified that: "Other departments are particularly assistive and collaborate closely with the nurse . . ."

Although materiel services in the critical care environment was not a subject of research in the nursing literature, having the right supply available at the right place and time, in the right quantities, in the right condition, and at the right cost has implicit benefits to the patient.

Another dimension of the support systems sphere in the critical care environment is the area of high technology and automation. Computer-dependent monitoring devices and equipment are a given in the critical care units. Research related to patients' recollection of critical care suggests that patients often associate their experiences in the critical care unit with the technical aspects of the environment (13). Managerial considerations related to this aspect of the environment include the following: potential for clinician's dependence on technology, acquiring a false sense of security related to diagnostic parameters, potential increase in iatrogenic injury to patient and clinical staff, higher risk factors increasing hospital and clinician liability, increase in clinician stress, and potential for increase of patient and family anxiety (14). Conclusive data relating positive patient outcomes with invasive procedures of monitoring techniques are not found in the research in significant numbers.

Physical Sphere

The physical sphere is the very tangible part of the environment and encompasses the controllable dimensions that are responsive to physical and spatial manipulation. The objective physical characteristics of environments such as noise levels, shapes, sizes, and physical structures are part of the physical sphere. The physical aspects of the critical care environment play a major role in influencing quality of patient care delivery. It is at this level where "things-people" interface within the environment. This sphere is closest in the model to the sphere representing human resources and the critically ill patient. In the Magnet Hospital study (9), nurses commented about "the importance of the physical plant as it contributes to the hospital's being a good place to work." The study done by Kraegel et al. (15) demonstrated that carefully designed medication distribution, materiel distribution, food distribution, linen distribution, on-unit communication systems, and physical layout of the unit increased nurses' time for direct patient care approximately 50% after the change. Although patient and nurse satisfaction increased, the quality of care did not increase significantly in the indicators they measured (15).

Critical care environment is associated with high technology and innovative treatment. The risk of potential injury to the staff members and to the patient is extremely high. There is no doubt that the importance of safety practices within critical care cannot be underestimated. Safety practices generally deal with information and prevention. Standards of safety specifically addressing the critical care environment have been written by AACN and SCCM. Environmental management includes the ability of nurse managers to create a safe and functional work environment. AACN relates the importance of physical environment to critical care nursing practice in the book entitled, *AACN: Standards for Nursing Care of the Critically Ill* (16). The relationship of safety practices to patient care outcomes can usually be interpreted through inference in the literature rather than through direct data correlations. Studies relating safety practices to patient care outcomes and nursing procedures should be encouraged because of the continual changes in technology and treatment.

Human Resource Sphere

The human resource sphere is the most complex and the largest component of the environment, and it transcends all the other spheres in the environment. Human resources have an impact on every part of delivery of nursing services: roles and relationships, the critical care nurse, the extended patient, the family, and the critically ill patient.

The nurse manager's responsibility is to facilitate the delivery of nursing services. The AACN's position statement, Role Expectations for the Critical Care Manager, states that "management of a critical care environment presents a particular challenge because of the magnitude of resources utilized, the timeliness of nursing intervention that is required, the sophisticated technological and clinical interventions that are used, the depth of the data base necessary for decision making, and the degree of collaboration needed among disciplines." It is further stated, "the role of the critical care manager is defined as the coordination and integration of human and material resources necessary to care for a population of critically ill patients" (17).

Roles and relationships are important considerations in the human resource sphere, particularly because roles and relationships influence collaborative practice. Knaus et al. (8) explained that one factor that made a difference in mortality data in critical care units was that collaborative practice existed in the critical care unit that had a lower mortality rate. Alt-White et al. (18) studied collaboration, but because collaboration is only one of four interrelated variables, the impact of collaboration on patient care outcomes is not isolated. There is growing opinion that collaborative practice between physicians and nurses in the critical care environment is essential to positive patient care outcomes. Knaus et al. (8) and Mitchell et al. (7) emphasized the need for continued research in the area of collaborative practice in critical care. In 1983, the Board of Directors of the AACN and the Council of the SCCM commissioned a task force of physicians and nurses to identify generic principles of collaborative practice in the critical care work environment. Included in the roles and relationships within the critical care work environment are the other paraprofessional groups, particularly the respiratory therapists and monitoring technicians (19).

In a survey of almost 1800 critical care nurses, Claus and Bailey (20) identified interpersonal relationships as both the greatest source of stress and the second greatest source of satisfaction. It can be postulated that barriers to forming positive interpersonal relationships can have a negative impact on patient care outcomes. Collaboration in the environment starts simultaneously at the bedside and from the top down. "The interview data reveal that in the magnet hospitals good relationships between administration and the staff prevail, with a focus on collaboration in executing the hospital's mission: providing care to patients" (9).

Customer relations is another aspect of the human resource sphere. The principles behind customer relations are risk management. An experience actually can be designed for the patient and family of the critically ill patient by nurses manipulating those factors over which they have control. Nurses can use elements in the environment to structure certain parts of the experience for the patient and family. Communication, written information, waiting room atmosphere, and visiting hours are examples.

Critically Ill Patient

The critically ill patient has been defined by ACCN as a person "who is characterized by the presence of or being at high risk for developing life threatening problems" (21). Although it is the critical care nurse who coordinates and provides care, the nurse manager plays a significant role in facilitating support services, and in integrating and maintaining goals and standards related to the quality of nursing services.

Of interest is the impact of the critical care experience on the patient. Simpson et al. (22) reviewed 59 patients' recollections of their critical care experience within 24 and 48 hours after transfer from the unit. Among the findings of this study, pain and sleeplessness were the negative experiences cited most. The nurses were associated with technical care and alleviation of concerns, and physicians were more frequently associated with being providers of information about health status.

CONCLUSION

A conceptual model of the critical care environment provides the nurse manager with a tool to evaluate alternative decisions. The nurse manager's role in environmental management is focused on balancing the interrelated elements in the environment to support the achievement of positive patient care outcomes for the critically ill patient. As indicated by the void in the research literature, there is a need for nursing researchers and clinicians to study the interrelatedness of elements in the environment and the correlations, causes, and effects on patient care outcomes.

REFERENCES

1. Stevens B. The head nurse as manager. J Nurs Adm 1974;4:36–40.
2. Ganong JM, Ganong WL. Help for the head nurse. Chapel Hill: W.L. Ganong Co., 1975.
3. Simms L, Price S, Ervin N. Creating the environment for nursing practice. In: The professional practice of nursing administration. New York: Wiley Medical Publishers, 1985:57–67.
4. AACN position statement: Scope of critical care nursing practice. Newport Beach: American Association of Critical-Care Nurses, 1986.
5. Keith JM. Florence Nightingale: Statistician and consultant epidemiologist. Int Nurs Rev. 1988;35:147.
6. Williams MA. The physical environment and patient care. Annu Rev Nurs Res 1988;6:61–84.
7. Mitchell PH, Armstrong S, Simpson TF, Lentz M. American Association of Critical-Care Nurses Demonstration Project: Profile of excellence in critical care nursing. Heart Lung 1989;18:219–237.
8. Knaus WA, Draper EA, Wagner DP, Zimmerman JE. An evaluation of outcome from intensive care in major medical centers. Annu Intern Med 1986;104:410–418.
9. American Academy of Nursing. Magnet Hospitals: Attraction and Retention of Professional Nurses. Kansas City: American Nurses Association, 1983.
10. National Commission on Nursing. Summary report and recommendations. Chicago: American Hospital Association, The Hospital Research and Trust, 1983.
11. Henry B, O'Donnell JF, Pendergast JF, Moody LE, Hutchinson SA. Nursing administration research in hospitals and schools of nursing. J Nurs Adm 1988;18:28–31.
12. Society of Critical Care Medicine Task Force Committee. Recommendations for critical care unit design. Crit Care Med 1988;18:796–806.
13. Simpson T, Armstrong S, Mitchell P. Demonstration project: Patients' recollections of critical care. Heart Lung 1989;18:325–331.
14. Sinclair V. High technology in critical care: Implications for nursing's role and practice. Focus Crit Care 1988;15:36–41.
15. Kraegel J, Mousseau V, Goldsmith C, Arora R. Patient care systems. Philadelphia: Lippincott, 1974.
16. Sanford S, Disch JM. AACN: Standards for Nursing Care of the Critically Ill, 2nd Ed. San Mateo: Appleton & Lange, 1989.
17. AACN position statement: Role expectations for the critical care manager. Newport Beach: American Association of Critical-Care Nurses, 1986.
18. Alt-White AC, Charns M, Strayer R. Personal, organizational and managerial factors related to nurse-physician collaboration. Nurs Adm Q 1983;8:8–18.
19. The organization of human resources in critical care units. Focus Crit Care 1983;1:43–44.
20. Claus KE, Bailey JT. 1980 living with stress and promoting well being. St. Louis: Mosby, 1980.
21. AACN Position Statement: AACN's Definition of Critical Care Nursing. Newport Beach: American Association of Critical-Care Nurses, 1984.
22. Simpson T, Armstrong S, Mitchell P. Demonstration project: Patients' recollections of critical care. Heart Lung 1989;18:325–331.

Part II

STRATEGIC SPHERE

Chapter 2
External Environment: Forces and Trends Reshaping Critical Care Nursing

KAREN LOGSDON

Few nurses would dispute the fact that the hospital environment has become more intense and difficult to manage. Something has happened that has caused many nurses to believe that they are not able to accomplish all that they want to accomplish for patients and their families. What has happened beyond our hospital walls to upset our time-honored relationships and methods of delivering patient care so drastically? Who are the major stakeholders and how have they responded? What has and might be the effect of these external forces of change upon the critical care units and the nurses who practice in them?

MEDICARE AND OTHER PAYMENT SCHEMES

The health care system, and particularly the hospital, is being driven by budgetary policy rather than a national health policy. There is little doubt that new financing and payment schemes for inpatient care have resulted in a crisis for many hospitals. This is clearly seen with a review of the Medicare population, which generally accounts for the greatest percentage of patient days, discharges, and single source of revenue for the typical hospital.

Faced with rising health care costs, the increasing national debt, and the demands of defense, transportation, education, and other public policy issues, the Congress established the Medicare Prospective Payment System (PPS). Since October 1, 1983, the PPS has paid for most inpatient acute care services according to rates set prospectively for each of the 476 Diagnosis Related Groups (DRGs) into which patients can be classified. The new payment system was phased in over 4 years (1983–1987). During this time, a hospital's payment moved from a reimbursement of the respective hospital's costs of treating Medicare beneficiaries (cost reimbursement), to a reimbursement system based on Medicare's average cost-per-discharge with adjustments to reflect cost factors outside of the hospital's control.

A few institutional provider groups have been exempted from the PPS: long-term care institutions, rehabilitation hospitals, children's hospitals, psychiatric hospitals, and certain distinct rehabilitation or psychiatric units of acute care hospitals. In addition, small rural, isolated hospitals that meet specified criteria have been designated as "sole community providers," which qualifies them for special payment consideration.

After a 4-year phase-in transition period, PPS is paying most hospitals on the basis of Medicare's national average cost per discharge. This so-called "standardized amount" is adjusted by each DRG's relative resource utilization weight. This DRG "price" is then adjusted by an Area Wage Index (AWI) to reflect the hospital's labor market costs. Other payment adjustments include the costs of direct and indirect medical education, capital costs, and losses of hospitals that provide a disproportionate share of services to unsponsored, uninsured patients. An outlier payment policy provides partial relief for atypically costly cases—as determined either on a cost or length-of-stay basis. These outlier patients are frequently critical care patients.

Additionally, in 1984, Congress started imposing price controls on hospital outpatient services when it capped outpatient laboratory rates. In 1988, Medicare completed a 2-year process of limiting a portion of hospital ambulatory surgical services payments to the rates it paid to freestanding ambulatory surgery centers. Further, the fiscal 1988 budget law (OBRA-87) enacted new limits on reimbursement for outpatient radiology and other outpatient diagnostic services such as electrocardiograms (ECGs) and electroencephalgrams (EEGs).

The result has been rapidly declining Medicare profit margins. According to the Prospective Payment Assessment Commission (ProPAC), the national Medicare profit margins fell from 14.7% in fiscal year 1984 to 8.2% in fiscal year 1986 (1). A recent study by the Health and Human Services Department's Inspector General's Office reports Medicare profit margins dropped to 4.77% in fiscal year 1988 and that the percentage of hospitals making a profit on Medicare declined to 57% from 66%.

For many hospitals, Medicare's reduced revenue stream barely covers the costs of providing acute inpatient services. The aggregate analysis of operating margins masks substantial variations among hospitals. During the first year of PPS, 10% of United States' hospitals had margins of −5.9% or lower, and another 10% had margins of 23% or higher. By the third year of PPS, the lower 10% had margins of −18.5%, while the top 10% had margins of 19.4%.

The Inspector General's study also found margin profits and losses varied among different types of facilities, with urban and teaching hospitals generally showing positive margins and rural hospitals losing money on Medicare. Urban hospitals earned an average 5.71% profit, while rural hospitals lost an average of 4.66%. The average profit level for teaching hospitals was 7.64% while nonteaching hospitals earned an average of 1.85%.

In any averaging system like PPS, there are "winners" and "losers." The primary reason is that a system like PPS fails to adjust equitably for differences in severity of illness, case mix differences, area variations in nonlabor costs, and labor market definitions, of which California has an unusual number. Over the course of the transition (which was completed for most hospitals in November 1987), 43% of California hospitals lost over $371 million due to PPS's national redistributive effect. The state as a whole incurred a net loss of $109 million.

The above discussion of Medicare profit margins, although a bit tedious, is important background information if nurses are to appreciate the fiscal constraints on hospitals. Medicare accounts for the largest percentage of inpatient revenue and approximately 20% of the nation's total health tab making the federal government the single largest factor in the health care cost equation. Critical care services constitute a large percentage of the Medicare volume. Because of the fiscal ramifications, nurses need to know how their hospital is faring with the Medicare payment scheme.

Further compounding the problem is the federal Medicaid program. Severe underfunding is well documented, as illustrated by California's experience with "contracted" rates. In 1987, Medi-Cal (California's Medicaid program) paid California's community hospitals $1.7 billion which was 67% of what it actually cost to provide inpatient ($2.5 billion) and outpatient ($200 million) services to 1,385,000 recipients. In 1988, 55% of the patient days in California hospitals were from Medicare and Medi-Cal, yet these accounted for only 44% of the direct patient revenues.

Further complicating the issue and challenging nurse managers is the fact that costs for Medicare and other services continue to increase. Often expressed as a percentage of the Gross National Product (GNP), the health care sector's share of the total economy has risen to 11.1%, up from 7.7% only 15 years ago. The fastest health care sector growth, however, is not hospitals but physician's services, other professional services, program administrative costs, and net cost of private health insurance (2). The total payment to hospitals as a percentage of GNP has remained constant since 1982, at slightly over 4% (3).

The Inspector General's report also found that costs for Medicare services increased twice as much as revenues in 1988. The average cost per Medi-Cal case jumped 11%, while revenues increased only 5%. Considerable debate occurs over the reason for increased costs. The Inspector General attributes the increase to inefficiency. Hospitals cite increased demand for sophisticated services, medical inflation, health system failures, increased numbers of uninsured, an aging population, and labor shortages, to name a few. Nurses need to be able to respond to accusations such as those of the Inspector General, particularly critical care nurses who work in intense resource-consuming environments.

Most experts believe that it is a mistake for the federal government to focus only on Medicare margins. A ProPAC study of teaching hospitals found that, while some hospitals made a profit on Medicare, many were losing money on their private-pay patients. This point introduces the last concepts to be discussed regarding the financial and payment schemes faced by hospitals.

As much from a reaction to cost shifting to private payors as to anything else, managed care and selective contracting for privately insured patients is further changing and impacting the way hospitals get paid. Seeing increases of 20% in health benefit premiums, big business has become disenchanted with private insurers. Through several mechanisms, third party payors are now contracting with hospitals and demanding considerable price discounts.

This trend is illustrated by the growth in Preferred Provider Organizations (PPOs). PPOs are associations of physicians and hospitals that contract with employers and insurers to provide health care services on a negotiated fee-for-service basis. In 1982, there were 45 operational PPOs and in 1987, there were 535 operational PPOs in the United States (4).

In addition to the underfunding of government patients and third-party contracting, there are a growing number of patients without any health insurance. Approximately 37 million persons have no health insurance at all, no Medicare, no Medicaid, and no private insurance. About one-third of these are children. In California, for example, it is estimated that 20% of the population is without health insurance, and that 75% of those are employed or dependents of the employed uninsured (5).

Critical care nurses are aware that critically ill patients consume many hospital resources and that these resources are delivered equally to all patients without making any distinction between those patients who have insurance at the time of service and those who do not. Whatever resources it takes to restore patients to stability are mobilized. This should not change. The financing and payment environment, however, is forcing hospitals to explore every possibility to increase productivity and contain costs.

Before discussing the challenges to critical care nurse managers, other external forces for change should be briefly reviewed. An aging population, technology, labor market issues, and maldistribution of health care resources create further pressures on the hospital environment and the critical care unit.

AGING

The United States population is projected to increase by 9.2% between 1988 and 2000. The age groups growing the fastest during this period are likely to be the 45- to 59-year age group with a 45% increase and the 75+-year age group. In 1988, 12,902,000 citizens were 75 years or older. In the year 2000, this is projected to be 17,244,000 and in 2025, it is projected to be 25,583,000. Between 1988 and 2025, a 98.3% increase in the 75+-year age group is expected (6).

The elderly population requires an increased intensity of inpatient service and resources. Elderly people account for 30% of all hospital discharges, 33% of health care expenditures, and yet they compose only 12% of the population (7). It would not be contested that with extended age comes chronic illness and an increased demand for rehabilitation and home services. Elderly people also account for a disproportionate number of highly technical services; 80% of all pacemakers, 80% of interocular lenses, and 75% of all hip replacements are for elderly patients.

TECHNOLOGY

The impact of technological advancements is seen at the extreme ends of the medical continuum. As a result of increased technological sophistication, many procedures and treatments originally requiring hospitalization are now done on an outpatient basis. On the other hand, new medical technologies have also resulted in increased demands for highly skilled nurses to monitor patients and equipment, to manage the side effects and complications of technological care, and to provide emotional support to anxious patients and their families.

New technologies affect the cost of care, the average length of stay, the inpatient-outpatient mix, staffing needs, and other hospital services. Technological advances improve patient outcomes and extend patient lives. Technological advances also contribute to growth in national health care spending.

New drugs, devices, or techniques that make procedures safer or more likely to succeed are expensive to develop and can result in more frequent performance of those procedures. Technology also adds to health care costs by prolonging lives of patients who may then require ongoing medical care.

Considerable pressures are placed on hospitals and physicians to control the utilization of expensive technologies and limited hospital resources. The most critical and controllable services are radiology, laboratory, and critical care units where excessive resource consumption may easily exceed stringent DRG limits.

LABOR MARKET

Health care labor shortages are a critical concern. Both supply- and demand-side forces must be considered. These issues are well documented in the numerous publications addressing the current nursing shortage. The Secretary's Commission on Nursing Final Report (8) concluded that the current shortage of Registered Nurses (RNs) is primarily a result of increased demand as opposed to a contraction of supply. The report concludes that "it is reasonable to expect that the tightening of hospital budgets, the compression of the full array of nursing services into shorter lengths of stay, the increasing average inpatient severity of illness that followed in the wake of the Medicare Prospective Payment System (PPS), and the decline in nurse's relative wage rates all promoted increased substitution toward RNs." More research is needed, however, to determine what the relationship may be between hospital payment methodology and the demand for RNs. Certainly the other external factors discussed play some role in the increased demand for RNs.

Experts agree, however, that labor is the most important element of health care inflation. The PPS rate is adjusted based on several factors. Inflation, as measured by the Health Care Financing Administration (HCFA) market-based index, is one factor. Experts recently agreed that the most important component of the market basket is salaries (9). Because nurses compose the highest percentage of hospital workers, the nursing shortage is a key factor in labor inflation. Labor costs are and will continue to be a critical issue for hospitals. As the costs for labor increase, hospitals will analyze critically the value of that labor. Forces to substitute capital for labor, to increase productivity, and to design alternative patient care delivery systems will increase. Critical care nurses will be challenged specifically by these factors. Compounding the issue are rising concerns over shortage of other health care professionals including pharmacists, physical therapists, and nonphysician practitioners.

ACQUIRED IMMUNODEFICIENCY SYNDROME

Although currently not a leading cause of death, the Centers for Disease Control (CDC) predict deaths from acquired immunodeficiency syndrome (AIDS) will reach 54,000 in 1991. The number of diagnosed cases has risen from 271 in 1981 and prior years to 20,159 in 1987. Deaths in 1988 were reported to be 21,000. Persons with AIDS require above average amounts of nursing care, and costs of medical care for each victim of AIDS is estimated at $60,000–80,000 from diagnosis to death. These costs and care needs, coupled with profound social implications, and the inability of medical science to as yet determine the etiology of the syndrome, place AIDS as one major force of change in the health care system (10).

NEW HEALTH CARE SYSTEM

Given the facts that Americans spend significantly more on health care, in comparison to other countries, that they have growing numbers of citizens with limited access to any primary care and who lack insurance, that they have severe underfunding and inequitable payments to hospitals, that they have labor shortages; and that they have a health care system marked by a paradox of plenty and paucity, it is little wonder that many people are talking about reform.

How to redistribute our health care resources to ensure a basic level of access for all people is currently being debated on many levels. The Congress, various state legislatures, consumers, providers, and payors are all forwarding proposals. These proposals take two basic approaches: build incrementally on our current system, or completely transform health care delivery in a manner similar to the Canadian health care system. Leadership from health policy experts, providers, and payors in a unique public-private partnership is required.

QUALITY

Concern over costs, the realization of unequal access, and the resultant rationing have generated an interest in quality as well as reform. Numerous national research efforts, both federal and private, are currently studying quality and effectiveness, identifying clinical standards, and developing medical outcome criteria. Hospitals are redirecting their efforts and systems to monitor and improve medical care.

Buried in the emerging debate on quality is not quality per se, but value. In a recent Harvard Business Review article, Jeff Goldsmith identified two critical quality questions: Does the medical procedure benefit the patient, given the cost? Could the procedure be eliminated, and the expenditure saved, without compromising the medical outcome (11)? Applying the value concepts, however, may create further complexity. How will society define benefit, given limited resources, an aging population, and new technologies? For example, judging new technologies by value fails to recognize the incremental nature of medical advances. Many of the technologies we take for granted today were developed in stages and did not emerge overnight.

OTHER EXTERNAL FORCES

A discussion of external forces reshaping the hospital would be incomplete without at least mentioning the changing relationships between physicians and patients and between hospitals and nurses, the impact of medical malpractice and rising premiums; and the numerous bioethical dilemmas faced by patients, providers, and society.

Nurses need an appreciation of the environment in which they are practicing. H. L. Mencken said it well: "For every problem there is a solution that is simple, neat, and wrong." These are very complex times for hospitals and nurses. Sustained, cooperative, and collaborative attention to many issues is required. The actions of providers, policy makers, and payors suggest that this is occurring. The answers will not come overnight.

IMPACT ON CRITICAL CARE

The above forces have all combined to move traditional inpatient care into the outpatient setting, the home health care setting, and into ambulatory clinics. Hospitals have diversified their services in an attempt to capture and provide as much of these services as possible. The remaining inpatient mix is increasing in complexity, caused, in part, by the forces of technology and aging. It may not be long before the acute care hospital consists only of critical care beds dedicated to treating critical injuries and organ failures. Management of chronic illness and restoring the already compromised patient to improved functioning will occur outside the acute care hospital in surgery centers, clinics, homes, and skilled nursing facilities.

This scenario presents several challenges to critical care nurses. Long recognized for their technical and specialized skill, critical care nurses have had very little experience with case management and coordination of multiple external services and agencies. It is likely that critical care nurses will have to play a greater role in discharge planning as more patients move directly from the intensive care unit to outpatient care.

The proportionate increase of critically ill elderly patients also presents special problems. The unique challenge of chronic illness, drug therapy and multiple interactions, and the aging process requires special knowledge. Complicated family interactions may also increase with a greater percentage of elderly patients.

The challenge to understand and support the special family and patient coping mechanisms associated with critical illness may increase. The potential that more patients will be discharged while still incommunicative or not having reestablished their normal patterns may create a further sense of personal isolation between critical care nurses and their patients.

Highly intense, critical settings will require increased specialization. Increased specialization, however, may add to the current dissatisfaction that hospital-based nurses express regarding their incomplete practice. Nurses regularly complain about not being able to perform patient teaching or provide emotional support. The future hospital environment may make it even more difficult. Professional nurses of the future may need to work in both the hospital and outpatient setting to fulfill their needs as nurses. The work environment may force nurses to redefine their practice and expectations. One treatment environment may not provide the opportunity to fulfill the entire scope of nursing practice.

Critical care nurses must also rise to the challenge of maximizing their resources and working more intelligently. The nurse-patient ratios and other high costs associated with treating critically ill patients require nurses to design the most efficient delivery mechanisms possible. One example is a critical care unit in Texas. Faced with a shortage of qualified nurses, staff members looked for time-reducing strategies that would not sacrifice quality. After clinical research, nurses have implemented a simple, disposable, nurse-controlled medicated device to administer Class II narcotics. This effected an annual cost savings of $43,000 (13).

Automated clinical records, bedside computers, and software for nursing case management are innovations that are available today, but acceptance has been slow. Nurses should take the lead in researching these high-technology labor substitutes and determine their benefits in the critical care environment.

Just as physicians are being challenged to measure the effectiveness of their medical decisions and treatment plans, nurses must develop clinical protocols to help measure the effectiveness and outcomes of nursing care. Before the end of the 1990s, the Health Care Financing Administration (HCFA) will expand its quality research efforts to the practice of nursing. Nurses can begin these efforts now by managing for value.

As the debate over reform of the health care system proceeds, hospitals will remain challenged by economic pressures. Until a national health policy and the available resources are redistributed, hospitals and nurses must survive in an environment of narrowing cash flows, deteriorating credit, and critical scarcities of technical and professional staff.

The patient care delivery systems are being looked at in new ways; historical traditions, departments, and services must all be re-evaluated. Nothing is sacrosanct in today's hospital environment. Physicians and nurses can be catalysts for effective change if they understand the challenges and can relate resources employed to the severity of the patients' illnesses. Hospitals that develop alternative delivery and patient care systems in collaboration with their medical and nursing staffs, will be better prepared to survive the 1990s and to journey into a new era of health care.

REFERENCES

1. Prospective Payment Asessment Commission's Report to the Congress, June 1988.
2. American Medical Association. The environment of medicine. Chicago, American Medical Association, 1989.
3. California Health Care Issues 1989. California Association of Hospitals and Health Systems, 1989.
4. American Medical Care and Review Association: Directory of Preferred Provider Organizations and the Industry Report on PPO Development, Bethesda, MD, American Medical Care and Review Association; June 1987.
5. CAHHS Marketplace Task Force Report 1989.
6. Spencer G (ed). Projections of the population of the United States by age, sex, and race: 1983 to 2080. U.S. Dept of Commerce, Bureau of the Census, 1984.
7. U.S. Senate Special Committee on Aging Report; 1988.
8. Secretary's Commission on Nursing Final Report; Volume 1, December 1989.
9. Health Care in the 1990s: Forecasts by Top Analysts, Hospitals Magazine, July 20, 1989.
10. Scitovsky A, Rice D. Estimates of the direct and indirect costs of acquired immunodeficiency syndrome in the US, 1985, 1986, and 1991. Public Health Reports 1987;102(1):5–17.
11. Goldsmith J. A Radical prescription for hospitals, Harvard Business Review 1989;67:104–111.
12. Alternative delivery systems for controlled drugs in the ICU. Presented at the 1989 Annual Meeting of the Association for the Surgery Trauma, Chicago, 1989.

Chapter 3

Strategic Planning: Designing a Future

JOAN GYGAX SPICER, MARYANNE ROBINSON

The health care environment today gives new meaning to the traditional management cycles of planning, implementing, and evaluating. People, capital, and time are scarce resources; therefore, nurse managers cannot afford to commit these resources without a well-defined intent or plan, nor can they afford not to expend the resources necessary to advance both the evolution of nursing practice and the delivery of nursing services.

The manager must provide a vision for the future and must develop leadership skills necessary to release the potential which exists in staff nurses in order to make a vision of the future a reality. The future will come about regardless, but the nurse manager is the one who facilitates the designing of a desirable future of nursing practice and the delivery of nursing services. Too often it is thought that if what is undesirable in the work environment is deleted, the future will be better, but removing what is undesirable does not ensure that what remains is what is desired (1).

The first step in the management cycle is planning. Planning activities no longer take place in isolation, because the environment's impact on the delivery of health care has created the need for linking the planning activities to external environmental factors, internal environmental factors, and values of the leaders and members of the work groups in the organization.

TAKING THE PROCESS OF DESIGNING A FUTURE A STEP AT A TIME

The future is not static, but an abstract, dynamic, and ongoing state. The process of designing a future is presented in discrete steps, but in reality, planning occurs as an ongoing process with less discretely defined steps. In a competitive market environment, successful organizations do rely on an ongoing planning process involving all levels of the organization. The results of good planning may almost seem like luck when an organization achieves success in the marketplace; success being directly correlated with successful management of services, human resources, and capital resources.

Step 1: Planning to Plan

Nurse manager leadership skills influence how the planning process is facilitated. Team building, supporting, coaching, promoting risk taking, and sharing of power are some of the necessary skills. The planning process may be top-down, bottom-up, or a combination of both. Top-down may be used if the manager does not value staff nurse participation or does not have the skills to facilitate a bottom-up process; however, if the organization is at immediate risk of survival in the marketplace, the top-down process may be necessary. A bottom-up process takes time and requires a skillful manager. A planning process from the bottom-up results in a solid plan because employees participated in the designing of the plan and, therefore, implementing the plan is easier. Using a combination of top-down and bottom-up incorporates expertise from all levels.

The planning activities need to be seen as exciting, so exciting that nursing staff members want to be involved. In preparing for planning the nursing staff and manager must become current and knowledgeable about events of the social, economic, political, professional, and organizational spheres. Some ways to accomplish this are using journal clubs, sending personnel to conferences, inviting guests to speak at staff meetings, posting summaries of legislative activities, and encouraging staff members to participate in their professional organizations. Planning begins when awareness of events surrounding health care delivery, delivery of nursing services, the practice of nursing, and critical care nursing is heightened, and nurses start thinking, talking, and questioning about the future.

Outlining the planning process includes defining discussion points, advisory points, and decision points for staff nurse participation. The first work to be accomplished includes identifying values related to the professional practice of nursing and delivery of nursing services, developing a mission and philosophy statement for the critical care unit, identifying critical success factors for the department of nursing, outlining the threats and opportunities, and identifying strengths and weaknesses. These elements influence the development of strategies. The strategies are broad goal statements, which are future focused and the umbrella under which the decisions that have an impact on nursing practice and the delivery of nursing services in the critical care unit take place. Umbrella strategies are important because the umbrella provides a longer range focus for the many daily activities. Strategies take into account the forces of the external and internal environment. Figure 3.1 shows the relationship among the elements in the strategic planning process. The first part consists of elements from which strategies are developed: values, mission, and philosophy,

```
            ┌─── Values
            │
            ├─── Mission and Philosophy
            │
STRATEGIC ──┼─── Critical Success Factors
            │
            ├─── Opportunities and Threats
            │
            └─── Strengths and Weaknesses

              ┌─── Goals and Objectives
              │
OPERATIONAL ──┼─── Action
              │
              └─── Evaluation
```

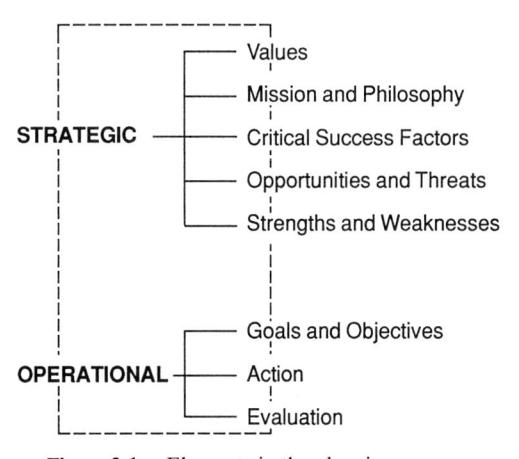

Figure 3.1. Elements in the planning process.

critical success factors, threats and opportunities, and strengths and weaknesses. The strategies usually encompass activities for the next 3–5 years. With the rapidly changing environment, the length of duration for the strategies is being shortened. The strategies direct the operational plan and the operational plan consists of the goals, objectives, and evaluation processes. Strategic planning is defined as "the process by which the guiding members of an organization envision its future and develop the necessary procedures and operations to achieve the future (2)."

Step 2: Professional Value Identification

Values are one of three bases from which a plan for the future is formulated. Professional values are integrated into a nurse's practice whether it be as a clinician, manager, or educator. Basic values are shared by nurses regardless of their role. The process of acquiring values is very complex. There are three steps in acquiring values: choosing, prizing, and acting. Values are freely chosen from among competing values after consideration is given to the alternatives. Values are prized and there is a willingness to make values known to others. Values are so integrated into the nurse's professional practice that the values drive behaviors. The professional values of each individual nurse may be different or whole groups of nurses may have homogeneous professional values. Examples of what nurses may value include: autonomy, collaborative relationships, and peer review.

As nurses form work groups, their values collectively create a culture. Culture is a set of values that members of the work group share that influence the quality of work performed. Culture is both an asset and liability. It is an asset because shared values ease and economize communications, and shared values generate higher levels of cooperation and commitment then would be otherwise possible. This is highly efficient. Culture becomes a liability when shared values are either not in keeping with the values of the organization or when they are not in the best interest of those receiving the service. Values drive norms. Sathe (3) defines norms as "standards of expected behavior, speech, and presentation of self." If performance excellence is a value, then certain work group norms can be identified. According to Alexander (4), some employee behaviors that support performance excellence are: trying to improve things even though the operation is running smoothly, discouraging other employees from the belief that doing only just enough to get by is not acceptable, and setting high personal standards of performance.

The point is that values drive the future directions of the organization from the top down starting with the Board of Directors and from the bottom up starting with the work group or staff nurse at the unit level. The more homogenous the values held by groups and individuals, the easier it is to design a future acceptable to all who are involved. When values are not homogenous, conflict arises and activities of the various groups and individuals within the organization are not productive in achieving organizational goals.

The nominal group technique can be used as a technique to identify which values the work group considers important. The advantages of the nominal group technique is that the focus is maintained on the task at hand and it protects against the phenomenon of a group process or discussion being dominated by a vocal few. Conducting the nominal process involves six steps. The task is presented to the group in writing and then read to the participants. Questions are answered and the task clarified. Step 1 is the silent generation of values in writing. Each member writes down what he or she thinks all nurses in the work group value. The second step is a round robin during which time one idea is taken serially from each group member. Each staff member lists one value while it is being recorded on a flip pad. Step 3 is a discussion to clarify the meaning of the items listed in the previous step. There is an opportunity to explore the logic behind the items. In step 4, the group members pick five items off the list and place each item in writing on a 3 × 5 card. Then they are asked to arrange the cards according to importance. Step 5 is the discussion of the preliminary ranking before the final ranking in step 6. At the end of this process, the facilitator and group members will have an idea of what the work group values.

The values and culture of the work group set the boundaries within which the planning process takes place and sets parameters within which the future can be designed. These values are one of three factors that influence the future directions of the practice of nursing in the patient care unit. "The degree of control that managers have over culture is very limited in comparison with the degree of control they have over structure, systems, and people's skills" (3). Culture may put a plan at risk.

The mission and philosophy is a brief statement of the critical care unit's services, patient population, financial goals, access to service, distribution channels, geographical areas serviced, and the driving values and beliefs. The values influence how the work is accomplished to achieve the mission of the organization.

A nursing department's mission and philosophy does not repeat statements or concepts found in the institution's mission statement. It is important that the critical care unit's mission and philosophy assist in implementing the organization's and nursing department's mission and philosophy. If outlined in a grid such as Table 3.1, the supporting statements from the unit level mission and philosophy can be easily identified. In each category, the nursing department's mission and philosophy has been outlined and the critical care unit's mission and philosophy have been aligned accordingly. Service as outlined for the critical care unit further defines the services of the nursing department as expert and specialized nursing care of the critically ill patient. The market to be serviced is specifically those patients with life-threatening conditions requiring critical care nursing. Desired market position of the nursing department is that of professionally recognized leadership in the development of nursing practice, and the critical care unit supports this by the nursing staff's desiring to be recognized both internally and externally for clinical expertise in critical care nursing. Financial goals for both the department of nursing and the critical care unit are to stay within allocated budget. Patients gain access to service within one of the multiple service settings of the department of nursing. Distribution channels vary for the nursing department, but the primary one for the critical care

Table 3.1. Grid of Mission and Philosophy Statement

	Department of Nursing	Critical Care Unit
Service	Expert and specialized professional nursing services	Expert and specialized nursing care of the critically ill patient
Market to be served	Patients requiring nursing services in a tertiary care center	Patients in life-threatening situations requiring care in a critical care unit
Desired market position	Professionally recognized leadership in the development of professional nursing services	Recognized clinical expertise and utilized as clinical resources for other patient care units
Financial goals	Develop, allocate, and administer resources with sensitivity consistent with corporate parameters	Stay within the allocated budget resources
Access to service	Within multiple practice settings	Within the critical care unit
Distribution channels	Administration, clinical practice, research, education, and consultation	Clinical practice and consultation within the nursing department
Geography	Selected regional, national, and international patient populations	Selected trauma and cardiovascular surgery patient population
Belief about patient	Humans are physical, psychological, social, and spiritual beings	
Belief about nursing	Nursing is the diagnosis and treatment of human responses to actual or potential health problems	Critical care nursing is the specialty within nursing that deals with human responses to life-threatening problems

unit level is clinical practice and consultation. Geography served is the same for both the nursing department and critical care unit. If specific patient populations in a high technology area are referred to the organization, the critical care unit may have unique geographical areas of service.

Beliefs about humans held by the members of the nursing department are that a human is a physical, psychological, social, and spiritual being. There is no need to repeat this at the unit level if the nurses in the critical care unit hold the same beliefs about humans. The nursing department accepts and believes that "nursing is the diagnosis and treatment of human responses to actual or potential health problems" (5). The critical care unit staff nurses accept the American Association of Critical-Care Nurses's definition of critical care nursing as being "the specialty within nursing which deals with human responses to life-threatening problems" (6).

Step 3: Scanning the External Environment

There are identifiable situations in the external environment that impact the critical care work environment whether it is in a small hospital in a rural community setting or a tertiary medical center in a cosmopolitan setting. These situations are beyond the control of the nurse manager, but nurse managers can evaluate how to utilize these situations in designing a future. When these situations can be used to an advantage, the situations are called an opportunity. Sometimes situations that exist in the external environment cannot be used to an advantage, but may cause potential barriers or constraints in designing a future; these situations are called threats. When planning takes into account the factors from the external environment, the planning process is commonly referred to as strategic planning.

The terminology surrounding planning may be confusing, but to clarify what is an opportunity and what is a threat, the question to ask is, "If I were a nurse or nurse manager in any hospital in the United States, would this situation have an impact on me, my work environment, or the patient?" If the answer is yes, then the situation is either an opportunity or threat. It is an opportunity if the situation can be used to an advantage and a threat if the situation can cause potential barriers or constraints in accomplishing future goals.

The urgency for the activities in the strategic planning processes is caused by the forces in the external environment. Among some of the actual forces requiring consideration are: reimbursement payment systems, population and health characteristics, available technology, available services, labor market characteristics, and supply of services and demands of service ratios.

Step 4: Critical Success Factors

The key to critical success factors is that these factors can be influenced by actions and decisions of nurses and nurse managers. Critical success factors are a limited number of areas in the practice of nursing and the delivery of nursing services in which, if performance is successful, the nursing department contributes to the competitive performance of the overall organization. These factors are derived from the external environment in a way similar to that of opportunities and threats. If one would ask the question, "If I were a nurse or nurse manager in any critical care unit in any size hospital, would it be critical to accomplish these factors? If so, are these factors ones over which I, as a nurse or nurse manager, have control?" If the answer is yes to both of these questions, it is most likely a critical success factor. Critical success factors are in the areas of marketing innovations, human resource organization, human resources, physical resources, productivity, social responsibility, and financial accountability (7). Proposed critical success factors for nursing services are:

- Marketing: Develop nursing care as a competitive edge for the organization.
- Innovation: Nursing management team provides expertise in program development and administration.
- Human organization: Thinking, learning, and creativity of employees is supported by management activities.
- Human resources: Highly qualified nursing care providers are available.
- Physical resources: Resources are allocated and structured to achieve economies of scale in patient care.
- Productivity: Delivery of quality nursing services is efficient and effective.
- Social responsibility: Nurses are active in influencing directions of social policy.
- Financial accountability: Delivery of services is cost effective (8).

Step 5: Strengths and Weaknesses

Strengths and weaknesses are identified after reviewing the internal environment. Organizational resources compose the internal environment, such as: management style, collaborative practice, organizational structure, patient care support systems, skill levels, technical ability, and financial position. When designing a future, there are certain resources within the organization that are strengths and there is also the lack of resources, limitations, faults, or defects that are weaknesses to consider in designing a future. Oftentimes, strengths and weaknesses are confused with opportunities and threats. The question that must be asked is, "If I were a nurse or nurse manager in any hospital in the United States, would these resources or lack of resources be the same?" If the answer is yes, then it is an opportunity or threat. If the answer is no, then the available resources can be referred to as strengths and the lack of resources as weaknesses. Strengths and weaknesses are organization- or patient care unit-specific. Strengths and weaknesses change from organization to organization and, within an organization, these may vary from patient care unit to patient care unit. Examples of strengths and weaknesses are presented in Table 3.2.

During the critical care unit planning process, the organization's, department of nursing's, and unit's specific strengths and weaknesses need to be identified from the perspective of the nurses upon whom the plan will have an impact and who are responsible for implementing it.

Table 3.2. Examples of strengths and weaknesses

Strength:	Clinical Nurse Specialists serve as role models and bring clinical expertise to the bedside.
Strength:	The initial structure of a strong recruitment and retention program are in place, including nursing recognition.
Strength:	Support and participation of physicians in interdisciplinary colaborative practice.
Strength/Weakness:	A unit-level patient education committee exists, but a full scope of cardiac patient education materials and preprocedure educational materials are not complete.
Strength/Weakness:	Standards of nursing practice have not been developed or adopted to govern all areas of practice in critical care.

In the examples listed in Table 3.2, the role of the Clinical Nurse Specialist is identified as a nursing organization strength. Clinical nurse specialists serve as role models and bring clinical expertise to the bedside. Another nursing organizational strength identified in this example is the initial structure of a strong recruitment and retention program being in place, including a nurse recognition program. This is another example of a strength of the nursing organization that has an impact on the critical care unit and is identified in the unit level plan. A strength within the critical care unit is the support and particpation of physicians in interdisciplinary collaborative practice. This is a specific strength in the critical care unit, and not all nursing units in the organization may have the same type of involvement by physicians in collaborative practice. A strength and weakness at the unit level is patient education. The strength would be the existence of the unit-level patient education committee, but the weakness would be the lack of complete cardiac and preprocedure educational materials. Standards of nursing practice not having been developed or adopted to govern all areas of practice in critical care is a unit-level strength and weakness. The strength would be that standards of nursing practice have either been developed or adopted, and the weakness would be that there are areas of clinical practice that are not governed by standards of nursing practice.

Strengths and weaknesses are the perceptions of those originating the plan. Other personnel within the organization may not perceive the same strengths and weaknesses. Within the planning process, the identification of strengths and weaknesses, if not handled appropriately, may put staff members at risk politically. There must be a trust within the working relationship of the group developing the plan and trust within the relationships between the nurse manager and the nurse executive in order for the strengths and weaknesses to be clearly delineated in an atmosphere that is conducive to honest expression of thoughts. Trust within the work group and between work groups is a necessary element because trust creates an environment for effective communication (9). The perceptions of staff nurses and the nurse manager of the strengths and weaknesses contribute to defining the boundaries within which the strategies and operational goals of the plan can be developed.

Step 6: Formulating Strategies

Strategies are broad goal statements that are future-directed and that encompass relationships between the critical care unit and its environments, both external and internal. Effective strategies identify desired future states and the means by which the desired states can be made to occur. The strategies guide the goals, objectives, and activities of the members of the nursing staff in the critical care unit.

In a bottom-up planning process, strategies are collectively developed by staff nurses, the nurse manager, and the nurse executive. Ohmar (10) describes strategists as having intellectual elasticity and flexibility. Not all people have the ability to strategize. This part of the planning process may be difficult to achieve, although if the external and internal environmental elements needed on which to base the strategies are carefully identified, the development of strategies can be easier, patterns will emerge, and glimpses of a future can be seen in one's mind. Strategies are broad future-directed statements, which means the statements are only general outlines. It will take 3–5 years to define the nature of the details within these general outlines.

Figure 3.2 conceptually represents the linkages with the environment, organizational culture, and internal environment. The external environment is examined during the planning process in order to identify the threats and opportunities. The concepts and beliefs that guide behavior within the organizational culture are reviewed and values of the work group are identified. The internal environment consists of the organizational resources: management style, collaborative practice, organizational structure, patient care support systems, skill levels, technical ability, and financial performance. The analysis of the internal environment yields the strengths and weaknesses. The mission is the fourth source of information. The strategist must evaluate information from these sources and formulate strategies that will minimize threats, maximize opportunities, build on strengths, minimize weaknesses, or change weaknesses, and also consider the values of the work group and leaders within the organization. Table 3.3 presents three examples of strategies. One example of a strategy addressed is shared governance. The strategy is to have a shared governance structure in the critical care unit. The operational goal is to increase the unit's readiness to implement a shared governance model of practice. Objectives to be accomplished in

Figure 3.2. Linkages with the environment. (Copyright 1987, Spicer and Spicer Associates, Santa Clara, CA.)

Table 3.3. Strategies

Strategy: Shared governance
    Goal: Increase the critical care unit's readiness to implement a shared governance model of practice.
    Objectives:
    1. Identify values and ground the critical care unit's mission and philosophy.
    2. Complete development and implementation of standards of practice.
    3. Develop unit-based quality assurance program.

Strategy: Recruitment
    Goal: Undertake activities that collectively provide a competitive edge in recruiting from a limited pool of available nurses.
    Objectives:
    1. Develop and market education and training programs.
    2. Develop an applicant-friendly system.

Strategy: Manage the perception of quality
    Goal: Educate the family on indicators of quality and identify the patient indicators of quality.
    Objectives:
    1. Interview patients post transfer from the critical care unit to obtain an evaluation of services and to identify what the patient describes as quality.
    2. Develop a family information brochure.

order to achieve the operational goal and, therefore, to move toward accomplishing the future strategy are identifying the nursing team's values and grounding the critical care unit's mission and philosophy statement in these values. Other objectives include completing the development and implementation of standards of practice and developing a unit-based quality assurance program.

The second strategy addresses recruitment. The strategy is to undertake activities that collectively provide a competitive edge in recruiting from a limited pool of available nurses. The objectives are to develop and market training programs and to develop an applicant-friendly system.

A third strategy is to manage the perception of quality. Quality means different things to different people. For the purposes of this strategy, quality is the subjective evaluation of the process, commitment, performance, and outcomes as the patient and family sees them. The goals to accomplish this strategy include identifying the client's indicators of quality, and educating the client on indicators on which quality should be evaluated. The objectives are to develop a patient posttransfer interview questionnaire and procedure and to develop a family information brochure.

Step 7: Goals and Objectives

Operationally, strategies are the conceptual umbrella under which the goals and objectives that guide the daily work of the nurse manager are developed (11). Goals and objectives direct activities that further define the details of the future. While the future is the focus, the current day-to-day activity evolves in small incremental steps toward the future.

Goals and objectives focus activity, set priorities, and provide direction. Resource allocation is based on the priority of goals and objectives. Without goals and objectives to provide direction, the work day passes and the urgency of the day drives the activity, but only to an unidentified end. Efforts are diluted if there are no goals and objectives to provide direction.

Goals and objectives provide direction in communication. The content of these goals and objectives is based on the following:

- Nurse within the profession (external environment);
- Nurse in the institution (internal environment);
- Nurse in the patient care unit (organizational culture);
- Nurse in the therapeutic patient relationship (personal values).

Although there may be three sets of goals and objectives at any one time driving the activity at the patient care level—the nursing manager's set, the staff nurse's set, and the set of those who are involved in the collective activity of patient care—all should support one another. The critical care unit-level goals and objectives give purpose to the staff nurse's work and convey a sense of mission. Once formulated and communicated, the goals and objectives form the core of the communication network (12).

In carrying out the objectives, it is important to understand that the objectives focus on what has been determined as important. Urgent means the activity must take place now. Figure 3.3 displays categories of objectives; the work performed falls somewhere in one of four categories:

1. Important and urgent;
2. Important, but not urgent;
3. Urgent, but not important;
4. Neither important nor urgent (13).

Nurse managers may spend most of their time on any one given day accomplishing urgent objectives, but it is only the important objectives that make a difference in getting to a desired future. Because of the urgency of some matters, the important things are left for last and sometimes never get addressed. When too much time is spent on the urgent and important, goals and objectives are not clearly defined in order to guide behavior or to refocus behavior, and the manager falls in an activity trap. "When you're so focused on what you're doing that you lose sight of the purpose for which you're doing it, you're in an activity trap" (13). Therefore, objectives must be clarified. To avoid the activity trap set up by the urgent demands of daily operations, the unit objectives should be written, measurable, realistic, and should have time-specific progress report times and deadlines. Figure 3.4 is an example of a work sheet that provides a structure to monitor achievement of goals and objectives.

Evaluation

Strategies, goals, and objectives should have tentative timelines for completion. The evaluation phase is one of the strongest influences in the strategic planning process because of its relevancy in the processing of information. That is, once strategies, goals, and objectives have been identified, it is important to measure the progress and the effectiveness of the plan. Time has a way of changing or influencing group values and organizational philosophy and mission because of uncontrollable factors partly external to the work group environment. The evaluation phase gauges the environmental information and measures the validity of strategies, goals, and objectives. It may be that a specific strategy is no longer current or applicable or that goals and objectives require enhancement.

SUMMARY

Strategies identify and provide the foci that guide choices made in the daily operations of the critical care unit. The strategies are grounded in the professional values of the nursing staff. When steps have been taken to

IMPORTANT

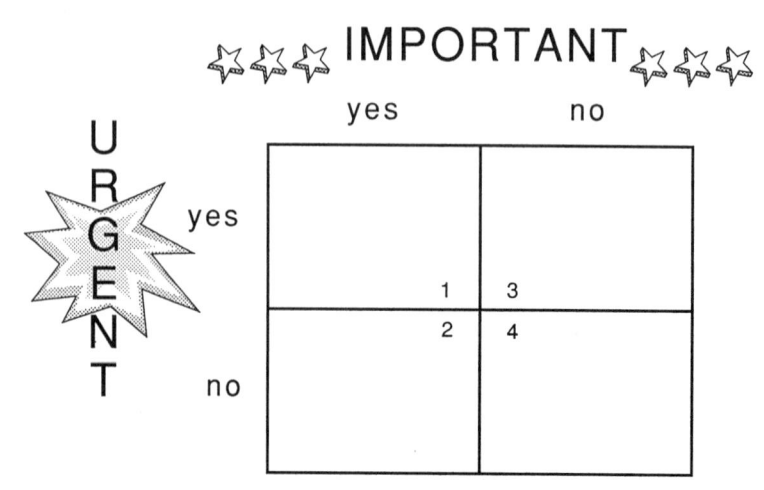

	yes	no
URGENT yes	1	3
URGENT no	2	4

Figure 3.3. Grid of urgent versus important. (From Douglass ME, Baker LD. How to develop clear-cut objectives and priorities. Chesterfield: Time Management Center, 1983;p 2.)

Strategy	Action Plan (Objective):	Responsible Person/s	Year 1 199x - 199x J A S O N D J F M A M J	Year 2 199x - 199x	Year 3 199x - 199x	Evaluation

Figure 3.4. Timetable work sheet.

understand the nurse manager's values and the staff nurse's work group's values, an important advantage can be gained in developing workable and well-supported strategies. If the strategies are not grounded in values of the staff nurse's work group, there will not be a personal commitment, nor will the strategy be implemented. A strategy in serious cultural trouble is likely to require some combination of three types of actions: managing around the culture, changing the culture, and modifying the strategy to bring the risk of failure into a manageable zone (3).

In reality, the structure of the document that encompasses the activity of planning is usually very brief and may be 15–20 pages at the most. The advantages of the plan's being documented are: establishing a sense of direction for the nursing personnel working in the critical care unit, communicating delegation of authority, describing the activities of the critical care unit in terms by which the interrelatedness of organizational activities can be evaluated in relationship to the commitment of resources, and providing an opportunity for other departments to contribute to the success of the critical care unit once the plan is shared.

REFERENCES

1. Ackoff RL. Creating the corporate future. New York: John Wiley & Sons, Inc. 1981:52–65.
2. Pfeiffer JW, Goodstein LD, Nolan TM. Shaping Strategic Planning: Frogs, Dragons, Bees, and Turkey Tails. Glenview: Professional Books Group, Scott, Foresman and Co., 1989.
3. Sathe V. Implications of corporate culture: A manager's guide to action. In: Organizational Dynamics: Corporate Culture. New York: American Management Association, 1988;5–23.
4. Alexander M. Organizational norms questionnaire. In Pfeiffer JW, Jones JE (eds). The 1978 Annual Handbook for Group Facilitators. San Diego: University Associates, 1978;83–88.

5. American Nurses Association. Nursing, A Social Policy Statement. Kansas City: American Nurses Association, 1980.
6. AACN Position Statement: AACN's Definition of Critical Care Nursing. Newport Beach: American Association of Critical-Care Nurses, 1984.
7. Leidecker JK, Bruno AV. Identifying and using critical success factors. Long Range Planning 1984;17:23–32.
8. Department of Nursing. Department of Nursing Strategic Plan: Into the Future. Irvine: University of California, Irvine Medical Center, 1987.
9. Johnson DW, Johnson FP. Cohesion, member needs, trust, and group norms. In: Joining Together: Group Theory and Group Skills, 2nd ed. New York: Prentice Hall, Inc., 1982;372–388.
10. Ohmar K. The mind of the strategist. In: Business Planning for Competitive Advantage. New York: Penguin Books, Ltd., 1982.
11. Strasen L. Key Business Skills for Nurse Managers. Philadelphia: J.B. Lippincott Co, 1987;201–203.
12. Spicer JG, Macioce VL. Retention: sound communications keep a critical care staff together. Nursing Management 1987;18:64A–64F.
13. Douglas ME, Baker LD. How to Develop Clearcut Objectives and Priorities. Chesterfield: Time Management Center, 1983.

Part III

ADMINISTRATIVE SPHERE

Chapter 4
Controlling Quality in the Critical Care Environment

MARYANNE ROBINSON, JOAN GYGAX SPICER

Nurse managers play a vital role in promoting and facilitating quality control activities at the unit level by creating an environment in which monitoring of quality of care is an integral part of the critical care unit functions, staff nurses are involved in implementing quality control, and collaborative decision-making is part of clinical practice. Controlling quality includes such activities as quality assurance, utilization control, and risk management.

Quality assurance (QA), utilization control, and risk management activities are not separate entities but inclusive of each other, because indicators used to monitor these activities in the environment are often based on the same standards. Indicators can be defined as measurable aspects of the quality and appropriateness of an important aspect of care or of service. Indicators can be derived from several sources in the environment. Integrating and linking these activities is essential when considering the impact of the results from these activities on patient care outcomes. In fact, the Joint Commission on Accreditation of Healthcare Organizations (JCAHO) recognizes the significance in the relationship between QA and risk management and intends to integrate the respective standards to assure coordination of activities (1).

Quality control activities can be described as systematic and planned sensors within a patient care environment. Figure 4.1 illustrates the relationship of quality control activities to the environment. These sensors act collectively like an environmental barometer by qualifying the climate or product, which is patient care.

Quality control activities consist of the following processes:

- Development and identification of critical indicators based on predetermined standards;
- Data collection methodologies;
- Data analysis and interpretation;
- Result reporting and discussion;
- Corrective action and follow-up.

Quality control activities are no longer considered a paper chase, but are a major focus of administrative directives as mandates and standards are enforced by regulating agencies.

QUALITY ASSURANCE

In the past two decades the concept of quality assurance has undergone evolutionary changes, although, historically, the nursing profession through Florence Nightingale, was the first to implement QA activities in its practice. QA has evolved from a problem-oriented approach, using retrospective medical records audits and usually conducted by a designated QA nurse, to a multidisciplined approach. The major thrust of these changes is a direct result of third party payor and consumer demands for quality care at a reasonable cost. JCAHO, in response to public demand, has played a leading role in redefining the meaning of QA.

QA is defined by most experts as the systematic monitoring and evaluation of the quality of patient care (2). The primary goal of QA is to maintain quality and provide where needed the mechanisms to improve patient care. In 1972, the enactment of the Social Security Amendment assigned the responsibility of evaluating the quality of care to the health care professionals accountable for providing that care (3). This policy change encouraged nursing services to adopt a decentralized approach to QA. Many definitions for nursing QA can be found in the literature; however,

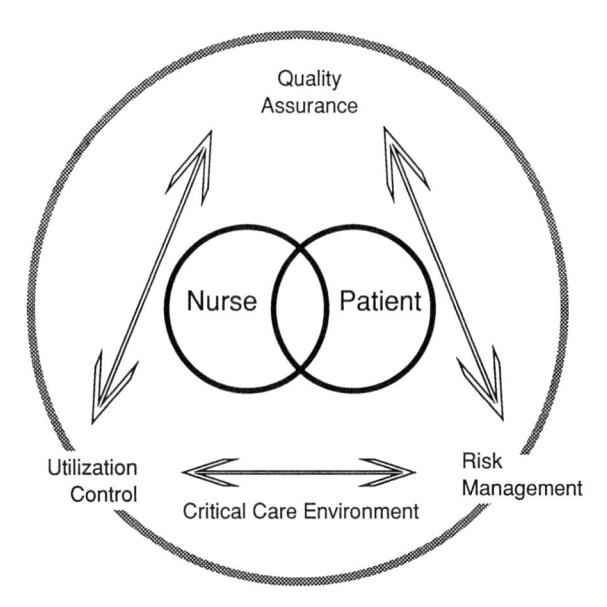

Figure 4.1. Relationship of quality control activities to the environment.

Gawlinski (4) best summarizes nursing QA as "a commitment to excellence in the delivery of nursing care, and a professional accountability for the care given." In addition to this definition, Maciorowski (5) provides three major goals of an effective nursing QA program. These goals include: (*a*) evidence of nursing accountability for services rendered and compliance with standards of practice; (*b*) a defined mechanism to identify, measure, and resolve clinical issues related to practice; and (*c*) a defined mechanism of evaluating quality indicators, collecting data, developing corrective action, and assessing outcomes.

In developing an effective nursing QA plan it is crucial that there is participatory involvement of clinical nurses in the development process. Active participation increases professional accountability, reduces resistance to change, and enhances goal achievement (6). A nursing QA plan provides the foundation and the framework of all quality control activities and outlines the direction of who, what, when, and how the activities are to be conducted. Included in the QA plan should be the following components:

- Clearly stated goal(s);
- Measurable objectives of how the goal(s) will be met;
- Designated accountability for written objectives;
- Delineated methods of QA activities;
- Outlined responsibilities conducting QA activities;
- Outlined mechanisms of reporting data;
- Outlined mechanisms of corrective action;
- Clear statement of confidentiality.

In addition to the above components, a QA plan for the critical care unit should be consistent with the overall hospital's QA plan as well as the nursing department's QA plan. The nurse manager provides the link between the unit and the other plans.

Because of the strong orientation to clinical aspects in the QA plan, it is recommended that clinical leadership should be provided by the clinical nurse specialist or senior staff nurse. In some critical care environments, the nurse manager may play a dual role of management and clinical expert. The role of the nurse manager is generally supportive and advisory.

Some of the primary objectives of the QA plan should include the development of unit standards, the development of standardized methods of reporting and communicating data, and the implementation of standardized data collection tools for problem identification. Once the QA plan has been developed, the responsibility of coordinating activities may ge given to a quality control committee; however, it should be stressed that participation of all nursing staff should be an expectation in achieving the goals of the QA plan. Participation in the monitoring of patient care should be part of the performance evaluation.

Nursing QA in the critical care environment is focused on functions of "any activity that directly or indirectly affects patient care" (4). Gawlinski (4) further outlines several sources in the critical care environment for evaluation, including nursing rounds, graphs, nursing flowsheets, education assessment, and peer-evaluated performance. In one article related to implementing QA in the critical care environment, the authors offer other sources for evaluation, such as incident reports, shift reports, patient care documentation, and problem analysis reports (7). In a recent press release related to QA, JCAHO provides important recommendations for the monitoring and evaluation processes of nursing services. They encourage QA activities including the following:

- Infection control;
- Drug usage evaluation;
- Pharmacy and therapeutics monitoring;
- Medical record review;
- Safety management.

They cite sources for monitoring activities in medication administration records, department logs, laboratory reports, meeting minutes, and patient and staff surveys (8). Reviewing these sources assists in identifying actual and potential problems so that monitoring activities can be prioritized.

The primary purpose of monitoring in nursing QA activities is to evaluate the nursing process and to identify actual or potential problems so that action can be taken to improve the quality of care delivered. There are three basic QA monitoring methods. Prospective monitoring identifies potential problems that may occur. An example would be monitoring safety practices related to use of a ventilator or an external pacemaker device. Concurrent monitoring evaluates actual compliance of patient care standards with current clinical or practice conditions. An example would be the monitoring of documentation practices on the critical care flowsheet. Retrospective monitoring reviews the delivery of patient care after the patient has been discharged. An example of this would be monitoring skin care protocol on trauma patients who required floatation beds during a certain timeframe in the critical care unit.

Monitoring activities should both be scheduled and should also be performed on an as-needed basis dependent on the unit's needs. A monitoring activity is usually initiated because either an actual or a potential problem was identified from environmental sources. Once the purpose of the monitoring activity has been clearly stated and outlined, standards related to the study should be identified. Standards are derived from either internal institutional policies, procedures and practice, or external professional standards such as American Association of Critical-Care Nurses' (AACN's) standards of practice; or from external healthcare agencies, such as JCAHO standards. The three types of standards derived from these sources are structure standards, which refer to how the environment is managed or to the requirements of operation; process standards, which refer to how clinicians practice or deliver care; and outcome standards, which refer to actual patient care results related to either delivered care or lack of delivered care.

A review of related factors and current information enhances the perspective of the study. Once these preliminary steps have been completed, the methodology of the study can be outlined. Methodology includes study, population, study purpose, timeframe of study, how the study is to be conducted, and who is going to conduct the study. The major emphasis of this step is tool development. The monitoring tool consists of indicators derived from environmental sources based on predetermined standards. It is important that the tool is objective and clear. JCAHO recommends that each indicator have a threshold for evaluation. The threshold is the point at which evaluation is initiated or, simply stated, the desired percentage of compliance of the indicator being audited. Depending on the indicator and the relative importance of the standard, thresholds may vary (8). For example, the threshold for compliance related to the indicator, "Documentation of patient response to pain management," may be desired and expected to be 100%. The threshold for compliance related to the indicator, "Patient's ability to demonstrate understanding of the purpose for TED hose," however, may be desired at 85%.

Staff participants should be inserviced on the purpose and methodology of the study before implementation. After all data have been collected, tabulation and recording of data are completed. Results are forwarded to the quality control committee for discussion and interpretation. Recommendations are drawn up from the results. Results, summary of discussion, and recommendations are forwarded to the nurse manager and appropriate institutional QA committee or department. Ultimately, the governing board has the authority and responsibility for quality of patient care and must review the overall hospital QA program, of which the nursing QA activities are one part.

The nurse manager has the responsibility to provide input and to draw conclusion from the results. The nurse manager must make decisions to support recommendations and/or to implement appropriate corrective action as well as to facilitate follow-up for further monitoring if necessary. Depending on the factors of the study results, the report is discussed at the interdisciplinary meeting to resolve collaborative issues. All QA monitoring activities should be reported to the nursing staff members for their input and recorded in staff meeting minutes.

The Critical Care Nursing Standards developed by the AACN provide the foundation as well as the accepted concepts of nursing process in the development of monitoring tools. Table 4.1 provides some common sources of indicators of nursing practice in the critical care unit. The format of the table utilizes structure, process, and outcome standards interrelating with the monitoring methods. Indicators are not intended to be conclusive but are offered as guidelines for implementing QA in the critical care environment.

UTILIZATION CONTROL

The major impetus for utilization control was spurred by federal regulations (Social Security Act, Title VI) regarding prospective payment, which stated, "hospitals (must) maintain an agreement with a utilization and quality control peer review organization . . . under which the organization will perform functions . . . the review of the validity of diagnostic information provided . . . the com-

Table 4.1. Indicators for Nursing Quality Assurance

	Standards		
	Structure	Process	Outcome
Methods			
Prospective	Monitoring equipment standards Alarm set standards Emergency cart inventory	Documentation standards on critical care flowsheet Patient education flowsheet	Patient's ability to demonstrate SQ injection according to prescribed method
Concurrent	Professional licensure CPR Certification Performance evaluation Physical environment safety inventory	Compliance to standards of practice of invasive procedure set-ups Compliance to standards of practice related to administration of vasopressor medication	Diastolic BP maintained <90 during vasopressor therapy
Retrospective	Educational assessments of professional staff	Documentation of execution of standardized protocols	Infection rate of patients with arterial puncture

pleteness of admissions and discharges and the appropriateness of care provided for which additional payments are sought." The primary goal of this policy is to promote the provision of quality care and to maintain effective use of health care services.

Turley's and Edwardson's recent review of the literature (9) related to the effective use of critical care suggested that there is a lack of substantial documentation to justify the benefits of critical care, that there is nonexistent use of criteria-based admission and discharge standards by physicians, and that there is evidence that hospitalization costs could be reduced. Baggs' historical review (10) of development of the intensive care units corroborates Turley's and Edwardson's findings (9) related to poorly documented evidence of the cost-benefits of critical care services. Baggs (10) adds, however, that empirical studies have been determined by others to show that intensive care decreases mortality and morbidity.

A patient is generally admitted to a critical care unit for one of the following reasons: (a) an immediate need for life-support therapy; (b) a potential risk requiring life-saving treatment; (c) the need for comprehensive and intensive nursing care (11). Patient diagnosis is not necessarily a criterion for admission. Recent utilization studies have suggested that 15% of all admissions into intensive care are unnecessary (12). More often than not, admission criteria are left to the discretion of the admitting physician who determines the need for critical care services. Knaus et al. (13), in an earlier study of consecutive admissions into intensive care units, found that the significant reasons for 49% of the admissions into critical care were for close nursing care and observation and not for treatment. Moreover, Baggs (10) surmised that the increased patient admissions into ICU for comprehensive nursing care have perpetuated the belief that good nursing care can only be found in intensive care, which has added to the demand for ICU beds.

It has been reported that intensive care units account for 20% of all hospital costs (11). National health care expenditures for intensive care are estimated to be $.15 to the dollar or 20 billion dollars annually (12). In one study, ICU cost was currently estimated to be 1% of the gross national product (10). Considering the cost of human resources related to the nurse-patient ratio required in order to perform nursing care efficiently, the cost of implementing advanced technological and clinical practices, and the cost of continuing educational requirements for staff members working in a dynamic and changing environment, it is not surprising to see the increasing interest for utilization control in the critical care environment.

Utilization control can be defined as those quality control activities monitoring the utilization of human resources in relationship to the efficient and effective use of services and monitoring the appropriateness of service provided. The primary objective of utilization control is to prevent the unnecessary and inappropriate use of health care services (14). It should be noted that utilization review is only a part of the broad concept of utilization control. Review connotes a formal mechanism evaluating appropriateness of part or all aspects of hospitalization, usually performed by the auditing of medical records using criteria outlined by professional review organizations. Influenced by congressional concerns about the impact of cost-containment measures and access to care, professional review organizations have been obligated to measure clinical indicators related to under- and over-utilization of health care services (15). However, Marker (16) believes that utilization review activities address two other questions: "Does an area have adequate resources and are these resources being used properly?" Moreover, formal utilization control studies can also provide data to substantiate the benefits of intensive care and justify cost. Sebilia (12) believes that as utilization review activities increase, valuable information related to proper usage of critical care resources will contribute to substantiating the appropriateness of critical care services.

Utilization control involves monitoring and relating the following indicators to the quality of patient care:

- Human resource dollars (cost per patient day);
- Daily census;
- Patient days;

- Length of stay;
- Direct admissions, transfers in, transfers out, discharges;
- Hours of care (hours per patient day);
- Severity of illness;
- Intensity of service;
- Diagnosis;
- Mortality and morbidity;
- Iatrogenic complications;
- Nosocomial acquired infections;
- Patient demographics.

Methods of evaluating utilization include tracking of resources (human and material) and of dollars used to provide services, trending of unit statistics, monitoring characteristics of patient population, tracking of nursing care hours used to provide patient care services, and tracking adverse patient care outcomes. Collecting utilization control data should be formalized at the unit level with monthly reports summarizing the data. The use of a unit activity log book, problem analysis reporting forms, and patient classification reports are just a few common formal sources in data collection. Data should be analyzed and interpreted by the management team and both positive and adverse variables should be identified. A summary of findings and recommendations should be forwarded to the appropriate hospital administrator, department, and/or committee as outlined by institutional policy for review and communication.

Table 4.2 outlines the common indicators measuring utilization and appropriateness of services. The indicators outlined in this table have been abstracted from a review of the literature and are not intended to be conclusive because state, community, and institutional standards vary from one area to the next. The table is divided into a horizontal flow chart diagraming a possible theoretical relationship of the indicators to each other.

RISK MANAGEMENT

While the health care industry continues to operate in an era of scarce resources, nursing, as the largest provider of health care in the hospital setting, has had to play a leading role in cost-containment strategies. Hospitals have been under pressure to contain health care costs in an environment of increasing medical complexity and sophistication. Nursing has had to adapt to these changes while maintaining the quality of care. Risk management in the health care industry is a result of the medical liability crisis during the early 1970s, when malpractice insurance premiums increased dramatically because of the sudden increase in consumer claims. Considering these facts, the importance of risk management cannot be overemphasized.

Risk is defined as a potential for injury or loss. Risk management can, therefore, be defined as the ability to control and/or prevent the possibility of injury or loss. According to Greve [17], hospital risk management is defined as the "development and direction of strategies for preventing patient injury, minimizing financial loss, and preserving hospital assets." Generally, hospital risk management programs are concerned with professional and general liability issues. They have the primary responsibility to reduce and minimize financial loss to the hospital. However, risk management to the nurse manager involves assuring safe clinical practice at the unit level.

Several factors must be considered when

Table 4.2. Utilization Control Indicators

Patient Safety	Clinical Practice	Procedures	Technology
Staffing mix	Competence	Medication	Equipment
Use of restraints	Performance	Administration	Function
Seizure precautions	Licensure	Life-support procedures	Electrical safety
Use of bed rails	Documentation	Equipment use	Alarm set
Informed consent	Discharge planning	Invasive procedures	Equipment procurement
Patient education	DNR policy	Diagnostic procedures	
Satisfaction		Infection control	

applying these definitions to the critical care environment. Factors characteristic of the critical care environment include:

- High consumption of health care dollars;
- Concentrated area of high technology;
- Concentrated area of innovative clinical practices;
- High risk patient population due to nature of severity of illness;
- Potential for legal, ethical, and moral dilemma related to life-death issues;
- Concentrated area of invasive procedures and treatment;
- High visibility due to intense involvement of family and others.

In addition to these factors, nurses have acquired increased accountability for practice. Nurses are faced with making clinical judgments and decisions related to legal, ethical, and moral dilemmas. It is the nurse manager in the critical care unit who is responsible for interceding as the gatekeeper in facilitating successful strategies for managing risks. The nurse manager's role in risk management includes implementation of standards, coordination of data collection, monitoring of data, and communication of data.

As in utilization control, there are identifiable indicators related to risk management in the critical care environment. These indicators can be derived from potential patient care incidents leading to adverse outcomes, hence causing injury or loss. A patient care incident is defined as an occurrence or an event involving an error in practice having the potential to cause harm, injury, or loss to the patient. The literature reveals few documented studies on factors influencing incident rates specific to critical care. Wan and Shukla (18) believe that hospitals in the United States "exhibit high variation in incident rates," partly because there is lack of standard reporting mechanisms. Moreover, incident reporting is one area health care clinicians fail to complete because of a false fear of liability and punitive actions. It is an accepted understanding that more incidents occur than are reported. However, some of the noted indicators for potential risk for liability include errors in medication, errors in intravenous administration, patient falls and injuries, and inappropriate diagnostic and therapeutic interventions (18). Wan and Shukla (18) further suggest the need for more studies in this area. Abramson et al. (19), analyzed 145 reported adverse incidents resulting from 4720 admissions in a medical-surgical intensive care unit in the late 1970s. Their findings revealed that 64% of the incidents were related to human errors and that 37% were related to equipment malfunction. Of the human errors, 47% were associated with equipment. Additionally, death rates were higher in patients with reported incidents than in those patients without reported incidents (20). Mitchell et al. (20) reportedly used clinical outcome variables such as mortality, complications, and patient satisfaction surveys and correlated these scores to valued aspects of the organizational environment; however, there was no elaboration on the clinical outcomes as they related to incidents common to critical care.

In April 1989, professional review organizations implemented a grading system to evaluate the quality of medical care. The purpose of this system was to ensure quality of care focusing on preventing harm to patients and to the physician (15). Risk management indicators derived from these standards include the following:

- Discharge planning;
- Stability of the patient at discharge;
- Deaths;
- Nosocomial infections;
- Unscheduled return to surgery;
- Trauma occurring during hospitalization.

Methods of risk management involves development of a structured monitoring mechanism, implementation of preventative strategies or standards, and persistent surveillance of potential patient care incidents. Initially, potential patient care incidents should be investigated immediately after the occurrence. This is a common responsibility of the nurse manager. A thorough investigation should be performed to delineate the factors leading to the incident. Appropriate corrective action should be immediate. All potential patient care incidents should be reported to the hospital risk management department.

Besides patient incident reporting mechanisms, risk management data are derived from

QA audits associated with potential patient care incidents such as ventilatory care, titratable medication administration, blood transfusion, and use of monitoring equipment. QA audits are proactive and assist in identifying problem areas that may lead to practice error. Critical care standards are useful references in developing unit-specific monitor tools. Although ongoing monitors do not eliminate the possibility that incidents will occur, they are a highly recommended preventative strategy.

Staff should be involved in developing and implementing standards of safe practice. Their involvement in the monitoring process is imperative and beneficial because it promotes accountability of practice. Potential patient care incidents should be a discussion item at the meeting of the quality control committee responsible for implementing standards in order to set up further monitoring.

In addition to immediate corrective action and ongoing QA monitoring, potential patient care incidents should be discussed in the collaborative forum to maximize the effectiveness of the corrective action. Potential patient care incidents should be traced consistently with a trend report completed monthly. Patient care incident trend reports are helpful when discussing collaborative approaches to improve quality of care in the unit. All reports and trends should be reported to the risk management department for information or further discussion. Trending of potential patient care incidents coupled with utilization reports and QA audits provides an objective mechanism to measure overall performance of clinical practice at the unit level. Table 4.3 summarized some of the common and broad indicators related to potential patient care incidents in the critical care environment. These indicators were abstracted from the literature and are not meant to be conclusive, because standards vary among institutions. Patient harm, injury, or loss may result from lack of or failure to comply with established standards associated with any of these indicators.

CONTROLLING QUALITY AT THE UNIT LEVEL

To be an integral part of the unit's daily functioning, quality control activities have to be integrated into nursing practice at the bedside. Traditionally, quality control activities in the hospital were the responsibility of designated departments and or committees. This centralized approach removed the focus of activities from the unit and the nurse. In a decentralized approach, quality control activities are directed from the unit level, where accountability for practice exists. The nurse manager's role in the decentralized approach includes directing, planning, implementing, and facilitating the processes of the quality control activities.

Staff involvement in quality control activities are inherent in a decentralized unit-based approach. The staff nurse provides the "experience" data in the development of current standards of practice. Staff members also assist in data collection processes. Upon anal-

Table 4.3. Indicators Related to Potential Patient Care Incidents

Population Characteristics	Unit Statisitics	Nursing Resources	Adverse Outcomes

Demographics
  Age
  Diagnosis
  Severity of illness
	Census
	
	Patient days
	Length of stay
		Cost per patient day
		
		Hours per patient day
			Deaths
			
			Iatrogenic
			  complications

Intensity of service
  Invasive procedures
 	
Noninvasive procedures
	Admissions
	  Direct
	  Indirect
	Discharges
	  Direct
	  Indirect
			Nosocomial Infections
			
			Cardiopulmonary arrest

University of California Irvine Medical Center

CRITICIAL CARE UNITS

QUALITY ASSURANCE REPORT

UNIT:_____   MONTH: _____   YEAR: _____

	Current Month	Patient Days	Previous Month	Patient Days
1. Number of Admissions				
2. Number of Deaths				
3. Number of Readmissions (Within 72 hours)				
4. Medication Errors				

5. **Deaths by Diagnosis and Cause:**

Patient File #	Diagnosis	Cause	Case Reviewed

6. **Major changes in medical management in the unit:**

Figure 4.2. Critical care quality assurance reports. (From the University of California, Irvine Medical Center, Orange, CA, 1987.)

ysis and interpretation of data, the staff members are key to the successful implementation of corrective action processes. Clinical participation in the planning stages of identification of the need for unit-based approaches toward quality control facilitates goal achievement and professional accountability. A mechanism to facilitate clinical participation is the implementation of a unit-based quality control committee designed to coordinate activities within the unit and to provide a communication link among peers and with the management team. The nurse manager plays an advisory and supportive role, while a clinical nurse assumes the clinical leadership in this committee. Ideally, this person should be a clinical nurse specialist who would enhance the QA process through clinical expertise. Members of a unit-

7. Patient Care Incident:

EVENT	YES	NO	EXPLANATION
Appropriateness of Admission			
Timeliness of Discharge			
Physician Response Time			
Coordination Among Services			
Safety-Hazardous Procedures			
Standing Orders / Actual Practice			
Respiratory Therapy			
Pacemaker Insertions			
Responses to Codes			
Infection Control			
Intravenous Therapy			
Patient Satisfaction			
Staff Complaints			
Line Insertions			
Tube Insertions			
Other			

_____ _____
Medical Director   Nurse Manager

Note: This information is protected form discovery and subpoena by California Evidence Code 1156 and 1157.

based quality control committee should include representation from each shift. Membership should be at least a 1-year commitment. Structured meetings should be mandatory and held at least once a month.

Collaborative practice among medical and nursing professionals is a crucial part of the decentralized approach. Collaborative practice in quality control activities promotes a unique forum in developing problem resolutions and corrective actions. The results reporting mechanism must be clearly delineated to ensure that appropriate actions are taken to improve the quality of care. Collaborative processes can be facilitated through monthly meetings between medical and nursing personnel of the unit. Members of this meeting should include the physician director of the unit, the nurse manager, and the clinical nurse specialists at the minimum. The purpose of this forum is to discuss the results from the quality control activities, such as patient care incidents, complications, or unexpected deaths. Corrective action and follow-up is also part of the purpose of these meetings. Although discussions are confidential in nature, meeting minutes should be recorded. Figure 4.2 is an example of a monthly report used in one institution that uses this approach.

The nurse manager is central to the success of a unit-based approach to quality control through a leadership role and the ability to utilize the results of the quality control activities to promote or improve patient care. Finally, it should be stressed that a unit-based approach does not minimize the need for or importance of a centralized structure, but rather it enhances the overall effectiveness of the QA program. Figure 4.3 illustrates a theoretical model of the decentralized and centralized structures.

Application of the unit-based approach to quality control requires a review and understanding of the principles of QA, utilization control, and risk management. Included in this understanding is the professional responsibility of confidential and ethical commitment regarding the operational aspects of quality control activities at the unit level. Planning the mechanisms, structuring the communication

QUALITY CONTROL ACTIVITIES

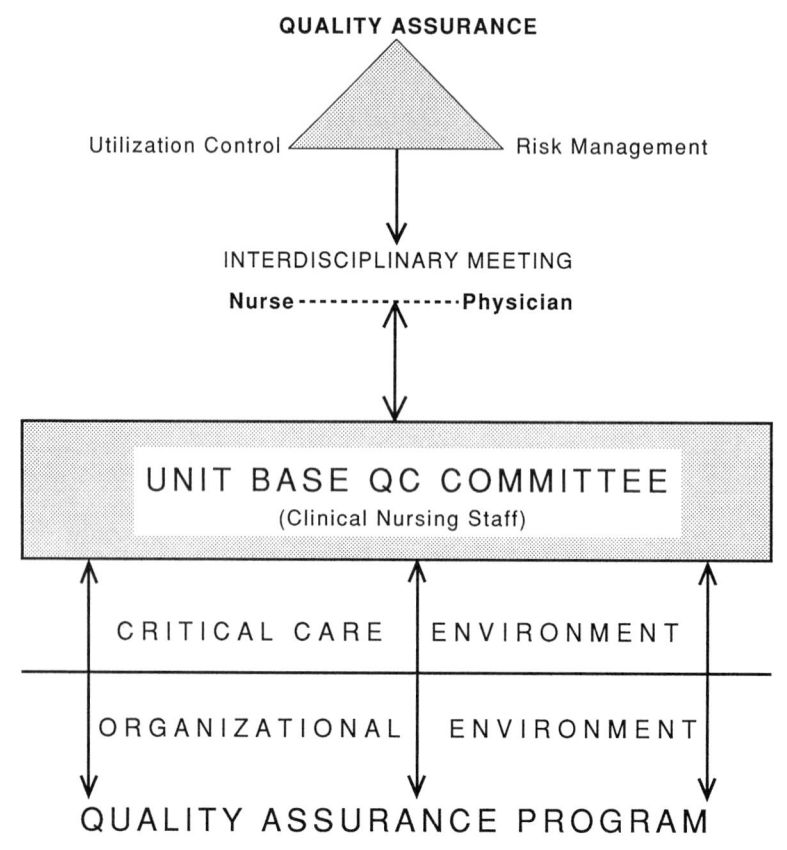

Figure 4.3. Application of quality control activities using unit-based approach.

links, and educating all participants before implementation are essential.

Finally, significant benefits in implementing unit-based quality control activities in the critical care environment include:

- Increased nurse manager's ability to identify in a more efficient manner problem areas regarding operational issues;
- Increased nurse manager's ability to intervene proactively and to predict future needs of the unit;
- Increased professional accountability promoting positive patient care outcomes;
- Heightened staff awareness regarding the importance of balancing quality and cost;
- Provision of the foundation for scientifically based clinical research and evaluation processes to occur at the clinical level where delivery of care is provided.

CONCLUSION

The complexities of the critical care environment demand attention to quality control. It is the nurse manager who holds the balance of quality control activities and effectiveness in the delivery of nursing care and the functioning of the critical care unit. The structure of the quality control activities facilitates the staff nurses' being able to meet their professional responsibilities to evaluate the nursing care that is delivered.

REFERENCES

1. Joint Commission of Accreditation of Hospitals. Risk Management: New changes. Chicago: JCAHO, 1986:1–2.
2. Scemons D. Quality Assurance Seminar. Los Angeles: Learning Tree, Inc. 1989.
3. An Act to Amend the Social Security Act, and for other purposes. PL 92–603, 92nd Congress HR-1.
4. Gawlinski A. Quality assurance and standards of care in the critical care setting. Crit Care Nurs Q 1982;6:43–49.
5. Maciorowski LA. Quality assurance. J Nurs Adm 1985;6:38–42.
6. Stahl LD. Demystifying critical care management. J Nurs Adm 1985;15:14–21.
7. Cottone D, Link MK. Quality assurance in critical care. Crit Care Nurs Q 1985;5:46–49.
8. Joint Commission of Accreditation of Health Care Organizations. Chicago: JCAHO, 1989;41.
9. Turley RM, Edwardson SR. Can ICU's be more efficiently used? J Nurs Adm 1985;15:25–28.
10. Baggs JG. Intensive care unit and collaboration between nurses and physicians. Heart Lung 1989;18:332–338.
11. Knaus WA, Draper EA, Wagner DP. Toward quality review in intensive care. QRB 1983;7:196–204.
12. Sebilia AJ. Critical care units. Crit Care Nurse 1984;2:36–37.
13. Knaus WA, Wagner DP, Draper EA, Lawrence DE, Zimmerman JE. The range of intensive care services today. JAMA 1981;246:2711–2716.
14. DiPrete RS. Impact of capitated finance systems on the continuum of care. Managing the continuum of care. Rockville: Aspen Publishers, Inc: 1987, 225–242.
15. PRO quality grading unveiled. Medical Executive Committee Reporter 1989:6–7.
16. Marker CB. The marker umbrella model for quality assurance. J Nrsg QA 1987;1:52–63.
17. Greve PA. Hospital risk management: Challenge of the '80s. Perspectives in Hospital Management 1986;6:9.
18. Wan TH, Shukla RK. Contextual and organizational correlates of the quality of hospital nursing care. QRB 1987:61–64.
19. Abramson NS, Ward KS, Grenvik AN, Robinson D, Synder JV. Adverse occurrences in intensive care units. JAMA 1980;244:1582–1584.
20. Mitchell PH, Armstrong S, Simpson TF, Lentz M. Profile in excellence in critical care nursing. Heart Lung 1989;18:219–237.

Chapter 5

Infection Control

BEVERLY A. POST
LYNN KREUTZER-BARAGLIA

One of the reasons for admission into the critical care unit is to use life-saving measures in the hopes of reducing morbidity and mortality in severely ill patients. Typically, critical care patients are more immunocompromised than those patients in general medical-surgical units. In uncomplicated cases, mortality rates may exceed 25% (1). However, more than one-third of the patients admitted to a critical care unit develop some sort of complication, the most frequent being nosocomial infections. Factors such as age and multisystem involvement increase risks for complications. In such cases, the mortality rate may exceed 40% (1). About 5% of all hospital admissions in the United States result in nosocomial infection (2). Statistics from the Centers for Disease Control Study on the Efficacy of Nosocomial Infection Control (SENIC Project) in the 1970s demonstrated that the national nosocomial infection rate was approximately 5.7 infections per 100 patients admitted or 2.1 million nosocomial infections each year. The cost to patients, institutions, and insurers has been estimated at more than $2 billion dollars a year (2,3). In an attempt to solve the problem of nosocomial infections, infection control programs have emerged.

Nosocomial infections are infections that develop during a hospital stay and generally are caused by organisms acquired during hospitalization. Nosocomial infection differs from community-acquired infection because incubation periods occur at the time of the patient admission. Infection manifests itself during the first 48 hours of hospitalization. A variable to consider in the identification of nosocomial infection is colonization. Colonization is the presence of a microorganism in or on the host that multiplies without the presence of any overt clinical signs of infection. In the absence of clinical signs, positive cultures should not be mistaken for infection.

CHAIN OF INFECTION

The chain of infection must be complete. For an infection to occur, three interrelated factors must be present: a susceptible host, an agent, and a mode of transmission. In addition, many factors must interact with the agent host to cause disease, thus explaining why everyone exposed to an infectious agent does not necessarily develop disease.

In an effort to control nosocomial infections, interruption of the chain of infection at its weakest link is considered the most effective intervention. By identifying the links in the chain for each nosocomial infection, the nurse should be able to initiate control and preventative measures.

Host Factors

The immune system plays a pivotal role in the patient's susceptibility to nosocomial infection. This is best demonstrated by examining the effects of immunodeficiency. Immunodeficiency, inherited or acquired, is a deficiency of the host's immune mechanism which leads to increased susceptibility to infection.

Inherited immunodeficiency diseases produce a variety of defects in humoral and/or in cell-mediated immunity. Acquired immunodeficiency diseases result from a plethora of causes including protein-calorie malnutrition, catabolic states such as cancer, chronic liver and kidney failure, viruses, and drugs. The acquired immune deficiency syndrome (AIDS) perhaps best demonstrates the sequelae of immunodeficiency.

AIDS, caused by the human immunodeficiency virus (HIV), is transmitted in semen, vaginal secretions, mother's milk, amniotic fluid, cerebrospinal fluid, blood, and blood products. Once transmitted, HIV attacks the T_4 helper cell of the host's immune system. The T_4 helper cell is the lymphocyte responsible for the immunological response to intracellular microbial agents and cellular antigens. HIV enters the T_4 cell by binding to a receptor on the surface of the cell. Once in the nucleus of the cell, the viral RNA of the retrovirus is transformed into DNA by the enzyme, reverse transcriptase. The transformed viral DNA integrates with the host DNA. Upon T_4 cell activation, the viral DNA produces more viruses. The active replication of the virus results in the death of the T cell and the release of the newly produced viruses. The ultimate result is a severe depletion of the T_4 helper cells.

To determine HIV infection, an enzyme-linked immunosorbent assay (ELISA) is performed to detect the host's antibodies to HIV. Because this test picks up antibodies to other antigens, it is necessary to follow a positive ELISA with a second test, the Western blot. If it is also positive, the individual is considered infected with the virus. False-negatives can occur with the ELISA because it takes from 6 weeks to 6 months for the body to produce enough antibodies to be detected.

The incubation period for AIDS can be as short as several months to more than 10 years. To assess progression from HIV infection to full-blown AIDS, T cell counts are done. HIV infection most characteristically reveals a depletion of T_4 helper cells with an inverted T_4–T_8 (suppressor cell) ratio. At this time, the most reliable single lab test to assess the immune status of an HIV-infected individual is the absolute number of T_4 cells. When T_4 cells number greater than or equal to 2 standard deviations below the mean, the subsequent immune dysfunction results in cutaneous anergy, lymphopenia, and susceptibility to opportunistic neoplasms, and parasitical, fungal, and viral infections.

The patient admitted into the critical care unit is admitted because of the severity of illness. Although the risk of nosocomial infection has been correlated to the degree of severity of illness, poor nutrition, invasive treatment, monitoring devices, and a variety of medications may further compromise the host defense mechanisms in the critical care patient.

This condition of having a high degree of host compromise in the critical care patient makes him more susceptible to infections. Therefore, it is necessary to assess how the underlying disease may have an impact on the patient's defense mechanism.

Because critical care patients are often ordered to have "nothing by mouth" and are in catabolic states because of underlying disease, nutritional status may be poor. Proper functioning of the immune system is dependent upon adequate nutrition. It is essential to undertake nutritional assessment upon admission to the critical care unit so that appropriate nutritional intervention can be planned to prevent the risk of infection.

Critical care treatment often involves the use of invasive treatment and monitoring devices. Vascular cannulas penetrate the skin barrier and provide a direct portal of entry into the blood. Intubation into the respiratory tract and/or Foley catheter insertion into the urinary tract bypasses normal anatomical defense structures. In addition, cannulas inserted into sterile cavities, such as the intracranial or the peritoneal cavity, expose the patient to pathogens.

Drug therapy further compromises the patient's defense mechanisms. An example is the use of immunosuppressant drugs directly affecting the host defense mechanisms. Drugs such as antibiotic therapy also interfere with hose defenses by destroying normal bacterial flora. For example, cimetidine alters the bacterial flora of the stomach.

In summary, host defenses are compromised not only by underlying disease, but by the course of treatment in the critical care unit. Careful assessment of each patient is necessary to determine the degree of host compromise.

Agent Factors

The second link in the chain of infection is the agent. The microbial agent in most nosocomial infections is a bacterium or a virus, occasionally a fungus, and rarely a parasite. Fungal and parasitical infections are seldom manifested in persons with essentially intact immune systems. Agent factors that may influence the chain of infection are the pathogenicity, specifically the virulence and the invasiveness of the organism, the dose, the antigenic make-up, and resistance factors of the organisms.

Pathogenicity is the ability of a specific organism to cause disease. High pathogenicity in an organism almost always causes disease in the host (e.g., *Shigella* organisms). An organism with low pathogenicity is *Staphylococcus epidermidis*, it commonly colonizes the host, rarely causing disease. Pathogenicity is further defined by describing the organism's virulence and invasiveness.

Virulence is the severity of disease caused by the organism. The severity of a specific disease can be determined by examining the morbidity and mortality rates for the disease. The greater the ability of the organism to invade tissues of the host, the higher the degree of the organism's pathogenicity (e.g., the *Shigella* species is highly invasive, eliciting a response by invading the tissues of the gastrointestinal tract).

Dose, another agent factor, is the number of organisms necessary for disease to occur. The number of organisms needed varies between organisms and hosts alike. For the *Salmonella* species known to cause disease in humans, a large inoculum of the organism is necessary to elicit disease in healthy individuals, whereas only a small inoculum of *Shigella* organisms needs to be ingested for disease to manifest itself in the same individual. The following formula expresses this relationship (2).

$$\frac{\text{Dose} \times \text{Pathogenicity}}{\text{Host Factors}} = \text{Infection}$$

Changing antigenic variations within certain species of microorganisms affects their abilities to produce disease. Such an example is the organism which causes influenza. Every 2–10 years a variant emerges which spreads rapidly because of the lack of host resistance to the new variant. This requires the development of a new vaccine.

The widespread use of antibiotics over the last 40 years has led to significant resistance in disease-producing microorganisms. One example of such is plasmid-mediated resistance. A plasmid in the staphylococcal organism enables it to produce a β-lactamase enzyme that inactivates penicillins and cephalosporins. Plasmid-mediated resistance can also be found in aminoglycoside-resistant enteric Gram-negative bacilli (e.g., *Pseudomonas*, *Neisseria*).

In both Gram-positive and Gram-negative microorganisms, multiple-resistant strains continue to appear with increasing frequency. Multiple resistance in bacteria is defined as resistance to two or more unrelated antibiotics to which the bacteria would normally be susceptible. In *Staphylococcus aureus*, a resistance to nafcillin, methicillin, and osacillin has given rise to a strain known as methicillin-resistant *Staphylococcus aureus* or more succinctly, MRSA. In Gram-positive bacilli, aminoglycoside (i.e., tobramycin, gentamicin, and amikacin) resistance is not uncommon.

Types of Agents

In the adult critical care, more than 50% of infections are caused by Gram-positive bacteria. *Escherichia coli*, *Klebsiella pneumoniae*, and *Pseudomonas aeruginosa* are the most common pathogens. The *Enterobacter* and *Serratia* species are less common but equally capable of causing life-threatening infections.

Gram-positive cocci (i.e., *Staphylococcus aureus*, group D enterococcus, *Staphylococcus epidermidis*) cause the remaining percentage of nosocomial infections in the critical care area. The *Staphylococcus* species is the prevalent organism in infections related to prostheses.

The following disease-causing agents or opportunistic pathogens cause disease in the immunosuppressed patient (e.g., the transplant patient, the patient undergoing antineoplastic therapy, and, more recently, the patient with AIDS): fungi (e.g., *Cryptococcus neoformans*, *Candida albicans*, *Candida tropicalis*, *Aspergillus fumigatus*, and *Aspergillus flavus*); protozoans (e.g., *Pneumocystis carinii*, *Cryptosporidium enteritis*, *Isapora belli*, *Toxoplasma gondii*, *Histoplasma capsulatum*); and viruses (e.g., herpes simplex, herpes zoster, cytomegalovirus, Epstein-Barr virus). Bacterial infections involve both common and unusual bacteria such as MRSA, *Mycobacterium tuberculosis*, *Mycobacterium ovum*, *Listeria monocytogenes*, and *Nocardia asteroides*.

Source and Reservoir

In addition to the agent factors, all organisms must have a reservoir and a source, which either may be the same or may be different. To be successful in the control and/or prevention of nosocomial infections by attacking the chain of infection, it is necessary to distinguish among these potentially different sites. A port of exit from the source and/or reservoir must also exist and be identified.

The source is the site from which the agent passes to the host. In the hospital setting, the patient, staff, or visitor may be the source of the infecting organism. If the infecting agent is from a source outside the patient, it is called exogenous. If the patient is himself the source for the infecting agent, however, it is called endogenous.

The reservoir, by definition, is the site where the organism survives to metabolize and reproduce. A reservoir can be human or animal (e.g., *Giardia*, *Staphylococcus*) or inanimate such as water (e.g., *Pseudomonas*) or food (e.g., *Staphylococcus*). As a general rule, viruses survive better in a human reservoir (e.g., HIV).

In the critical care setting, the inanimate environment provides many potential sources/reservoirs for infectious organisms. Invasive devices such as triple lumen catheters, pulmonary wedge catheters, arterial lines, intermittent access catheters, peritoneal and hemodialysis catheters, prostheses, implants, hyperalimentation solutions, Foley catheters, endotracheal tubes, chest tubes, and respiratory therapy equipment can be excellent sources of infection if improperly handled. In addition, stopcocks and pressure transducers are easily contaminated and capable of supporting the growth of microorganisms.

It is essential for the nurse to be able to identify the reservoirs or sources of infectious agents if control and/or preventative measures are to be used in breaking the chain of infection at this point. Any person entering the critical care unit is a potential reservoir/source of nosocomial pathogens. The nares or perineums of physicians and nurses may be colonized with and may shed *Staphylococcus aureus*, MRSA, or group A streptococcus. Active dermatitis can also shed large numbers of organisms.

Portal of Exit

An infectious agent's portal of exit from a human reservoir may be single (e.g., *Salmonella*: gastrointestinal tract) or multiple (e.g., MRSA: respiratory tract, urinary tract, draining wounds). Generally, the portals of exit are the gastrointestinal tract, the respiratory tract, the genitourinary tract, skin, and wounds. In addition, blood may be considered as a portal of exit as in hepatitis B and AIDS.

Mode of Transmission

The third link in the chain of infection is the mode of transmission. The mode of transmission is the manner in which organisms travel from the source to the host. The means or mode of transmission can be divided into four categories: contact (droplet or direct or indirect), airborne, vehicle, and vector routes.

Contact

Direct contact occurs when there is person-to-person spread. In this manner, organisms colonizing one person can be spread to cause infection in another person. Actively infectious material can also be transmitted by direct contact. Hands provide the most effective mechanism for direct contact transmission.

Indirect contact occurs when there is person-to-person spread. A person touches an inanimate object that has been contaminated by another person or inanimate object. An example of indirect contact spread is the transmission to susceptible hosts of respiratory tract pathogens on an inadequately sterilized bronchoscope initially contaminated when it came into contact with an infected patient. A break in technique during the insertion of a central line is another example. The bacterium, *Clostridium difficile*, produces stable spores that are thought to be passed via the indirect contact route and may explain the clusters of pseudomembranous colitis that can occur in critical care units.

Droplet spread refers to the passage of an

infectious agent through the air when the source of the agent and the perspective host are relatively near each other. Upper respiratory infections (e.g., strep throat, measles) are spread in this manner. When a person sneezes, coughs, or talks, droplets are disseminated that are relatively large in size (greater than 5 μs). They travel approximately 3 feet before settling on inanimate objects. Infections occur when the susceptible host in close proximity to the source inhales the droplets. Several respiratory viruses, including rhinoviruses, respiratory syncytial virus, and influenza viruses have entered critical care via droplet spread. Masks can help. Physical distance is not always an available option in the critical care setting. However, there should be a separation of at least 7 feet between critical care beds (1).

Airborne Transmission

The airborne route of transmission involves distances of more than several feet between the source and host. Particles from the upper respiratory tract less than 5 μ in size provide for airborne transmission of certain pathogens. These particles, from which all moisture evaporates before they fall, are called droplet nuclei and are small enough to get into the air currents and remain suspended. The tubercle bacillus can be spread in this way. Skin squamae can become airborne providing a means of transmission for *Staphylococcus* and, therefore, a source of infection in surgical wounds. Dust particles may settle on surfaces or may remain suspended for hours or even days. Sweeping, dry dusting, and shaking out of linen can cause organisms to become airborne. The mode of transmission for the organism responsible for Legionnaire's disease, *Legionella pneumophila*, is airborne. Water droplets from air conditioning cooling towers evaporate and are drawn into air intakes resulting in the dissemination of the bacteria. The *Aspergillus* species produces conidia that are inhaled following aerosolization.

Vehicle Transmission

Infection can be spread through contaminated blood, drugs, food, and water. Hepatitis non-A and non-B are still transmitted via contaminated blood transfusions. HIV was effectively spread through blood transfusions and blood products before the development of the ELISA test in 1985. Today, while not 100% free of HIV, blood transfusions and blood products are relatively safe.

Contaminated food has been associated with widespread outbreaks. Raw eggs, fruit, and vegetables have all been implicated in food-borne illnesses in immunosuppressed patients. An outbreak related to contaminated intravenous fluids is another example of vehicle transmission.

Vector Transmission

Vector transmission is not a common problem in the United States. An infection spread by the vector route has an animal or insect as its intermediate host between two persons. Plague and malaria are examples of the vector route of transmission.

INFECTIONS IN THE ADULT CRITICAL CARE UNIT

The most commonly occurring types of nosocomial infections in the adult critical care unit are lower respiratory tract infections (pneumonias), urinary tract infections, and bacteremias. Less frequently reported but prevalent among immunosuppressed patients are fungemias which are usually related to total parenteral nutrition, or treatment with antibiotics over a long period of time (1).

Nosocomial Pneumonia

Nosocomial pneumonias occur in up to 25% of critical care patients (1). The replacement of the normal flora of the upper respiratory tract by Gram-negative bacilli occurs in an estimated 50% of critical care patients (4). Subsequent infection is the result of the impairment of host defenses by any combination of the following conditions: obesity, advanced age, chronic obstructive lung disease (COLD), leukemia, coma, electrolyte imbalances, use of alcohol, and street and prescription drugs. Aspiration is another major cause of pneumonia in the critical care patient. Repeated aspiration of respiratory secretions colonized with Gram-negative bacilli can overwhelm the respiratory tract's natural defenses and result in pneumonia.

Iatrogenic factors further compromise the host's respiratory defense mechanisms. The most critical predisposing factor for nosocomial pneumonia is endotracheal intubation. The incidence rate for the development of pneumonia in intubated patients is estimated to be four times higher than that of nonintubated patients. The need for a tracheostomy increases the risk even further. The use of an endotracheal tube eliminates the natural host defenses of the upper respiratory system (i.e., the filtration system of the nose and airways and the mucociliary clearance system).

Respiratory therapy nebulizer equipment has long been known to be a source for nosocomial pneumonias. The institution of the cascade humidifier which allows the oxygen to bubble through water before delivery has greatly decreased the number of pneumonias seen in the critical care area.

Other iatrogenic causes of decreased effectiveness of critical care patients' defenses include suppression of the cough reflex by sedatives and narcotics, depression of mucociliary clearance by high oxygen concentrations, reduction of host immune defenses by drugs (e.g., corticosteroids, immunosuppressive agents), reduction of bacterial clearance by the macrophages due to pulmonary edema and/or high levels of oxygen concentration, the effect of antibiotics in the proliferation of more resistant strains of bacteria, use of antineoplastic drugs and radiation therapy in the treatment of malignancies, and surgeries.

Prevention of nosocomial pneumonias in the critical care unit is based on minimizing the colonization of the patient's upper respiratory tract and on decreasing transmission opportunities. This can be accomplished by frequent respiratory hygiene, careful attention to the care of respiratory equipment, isolation of patients, and handwashing.

Nosocomial Urinary Tract Infections

The urinary tract is the most common site for nosocomial infections. Although a urinary tract infection is not life-threatening in itself, it can lead to bacteremias, pylonephritis, and other sequelae that have significant morbidity and mortality for the critical care patient. The urinary tract defenses are compromised iatrogenically by the insertion of indwelling catheters which impair mechanical, mucosal, and voiding reflex actions, and by the application of condom-like catheters which allow urine to collect and stagnate, supporting colonization of the periurethral area. Colonization of the periurethral area with Gram-negative bacilli is a predictor of infection in patients with catheters. Nosocomial urinary tract infections are most prevalent among geriatric and critically ill patients. When catheters are present, scrupulous technique in the care of the catheter and the hygiene of the periurethral area are essential preventative measures.

Nosocomial Bacteremias

Of all nosocomial infections, bacteremia is the most dramatic and life-threatening. It can be a primary infection that occurs without the presence of any other infection caused by the same organism, or a secondary complication of an infection. Although only a few patients acquire a bacteremia, its association with significantly increased mortality warrants close monitoring in regard to potential of actual outbreaks. In one study involving 33 hospitals, one-third of all bacteremias occurred in the critical care unit (1).

In the critical care unit, intravascular devices are used for many purposes including hemodialysis, monitoring arterial and pulmonary pressures, providing large volumes of fluid, infusing medications and nutritional substances, obtaining blood samples, and inserting of pacemakers. These devices account for the majority of the iatrogenic-related bacteremias. They provide a continual source for the introduction of bacteria into the bloodstream. Drugs and nutritional factors compromise the critically ill patient. In addition, the critical care patient may have malfunctioning or actual loss of the organs that aid in the clearing of bacteria from the blood (e.g., patients with hepatic dysfunction or patients who have had splenectomies). Other factors involved in nosocomial bacteremias are surgical procedures, respiratory interventions, infected wounds, and contaminated intravenous fluids.

Infections In Implantable Prosthetic Sevices

Among the generally used implantable devices are cardiac valve prostheses, arterial grafts, permanent pacemakers, arteriovenous shunts, joints, and central nervous system (CNS) shunts. Unless there are indications other than the placement of the prosthesis itself, however, the patients most commonly seen in the critical care setting are those who have undergone heart valve implantation or the placement of CNS shunts.

Postoperative endocarditis occurs more frequently in patients who have undergone open heart surgeries. The use of heart-lung bypass machines seems to decrease the patient's natural defense mechanisms and increase the chance of intraoperative contamination. Infections present in other body sites at the time of surgery can act as sources for direct contamination of the operative site, predisposing the patient to bacteremia. This, in turn, can lead to the seeding of the prosthetic device.

The most common organisms associated with cardiovascular prosthesis infection are *S. aureus* and *S. epidermidis*. The most likely source for infection can be endogenous, from the patient's own flora, or exogenous, from operating room personnel who are shedders. Research has demonstrated the most important factor affecting the occurrence of postoperative infections in heart valve prosthesis is the use of antimicrobials before, during, and after the procedure (1). The extensive use of antibiotics before surgery has been identified as increasing the risk of infection in valve implantation. Their use is believed to cause replacement of the normal flora of the skin, mucous membranes, and gastrointestinal tract with antibiotic-resistant organisms. The prophylactic use of antibiotics during and after surgery has reportedly been responsible for the emergence of more antibiotic-resistant pathogens including various fungi. Therefore, prophylaxis with antibiotics should be limited to those circumstances where evidence has clearly demonstrated a benefit, or when the consequences of a surgical site infection would be catastrophic. A typical cardiovascular prosthetic infection is prosthetic valve endocarditis. The clinical indicators for prosthetic valve endocarditis are fever, changing murmur, and systemic embolization. Positive blood cultures will confirm the diagnosis. In addition, echocardiography may be useful in determining the presence of endocarditis.

CNS infection associated with the placement of a ventricular shunt is uncommon; however, it is the most serious of the iatrogenic infections. CNS infections are associated with a high mortality rate and with severe neurological sequelae. Infections associated with ventricular shunts are considered nosocomial when occurring within 60 days of implantation. It is estimated that 10–20% of patients undergoing shunt insertion develop infections as a result. Many ventricular shunt infections can be attributed to inadequate eradication of normal skin flora—especially *S. epidermidis*. The value of prophylactic antibiotic therapy is undecided. The clinical indicators for meningitis associated with ventricular shunt placement are no different from the indicators for any meningitis. They include fever, chills, headache, nausea, vomiting, cervical rigidity, and a positive Kernig's sign. In addition, there may be local erythema over the shunt course and shunt malfunction. Treatment may require removal of the shunt. The use of antimicrobials in the treatment of the infection is controversial related to the variability of CNS penetration.

Transplant Patient

A leading cause of death in organ transplant patients is infection. Any opportunistic pathogen can cause a serious infection in any post-transplant patient. Contributing factors in the development of infection include underlying disease, the site of implantation, and the immunosuppressive therapy regimen. The transplant group most susceptible to bacteremias is the bone marrow transplant group. Bacteremias also present a major problem for renal and liver transplant patients. Bacterial urinary tract infections are only a problem in renal transplant patients. Pneumonias are common to all transplant patients. CNS infections are a major cause of morbidity and mortality in the immunosuppressed patient. Incidence of bacteremia in transplant patients varies with the kind of transplant.

Bacteria are the most common agents associated with morbidity and mortality in transplant patients. This is especially true during the early post-transplant period. Gram-negative bacilli, usually enteric, are responsible for urinary tract infections, bacteremias, and pneumonias; Gram-positive cocci, however, are significant casual factors in patients requiring long-term intravenous therapy and indwelling central venous lines. *Listeria* remains a major pathogen in CNS infections and in bacteremias in renal transplant patients. Reactivation of hepatitis B infection originally contracted during pretransplant hemodialysis can be a serious problem for renal transplant patients.

Herpes simplex and zoster may also be reactivated after transplantation due to immunosuppressive therapy. While not life-threatening, reactivation of the herpes virus can be painful and can affect the patient's ability to eat. Fungal infections, specifically *Candida*, are common in bone marrow and liver transplant patients.

INFECTION CONTROL IN THE CRITICAL CARE ENVIRONMENT

Hospital-acquired infections are more frequent and are more easily transmitted among critical care patients than among other hospital patients. The essential components in the control of infection in the critical care environment include a strong infection control program, adequate staffing, and the structural design of the unit.

Design of the Critical Care Unit

The structural design of the critical care unit has a direct effect upon the potential for infection. Therefore, personnel who care for critically ill patients must be included in the planning and structural design of the unit construction or renovation.

The critical care unit should be located in the overall hospital plan in such a way that traffic can be controlled. Ideally, the unit should be in a cul-de-sac so that nonunit personnel and visitors are prevented from passing through the unit.

Today, single patient cubicles are considered state of the art. Physical separation lends itself to decreased spread of environmental flora as well as providing a barrier to cross-contamination. Separate cubicles also provide for respiratory or strict isolation of patients. Although reverse isolation for the immunosuppressed patient has not been proven as an effective means of preventing nosocomial infection, hospitals with open units should give serious consideration to placing a physical barrier between these patients. Standard guidelines dictate at least 120 square feet per cubicle (1). This allows adequate floor space for access to the patient and room for equipment. All surfaces should be smooth and constructed of easily disinfected materials. Each cubicle needs cabinets with doors in which to store patient care items. Ideally, each cubicle should have a ventilated anteroom to reduce airborne microorganisms and to provide a space for the storage of gowns, gloves, masks, and other equipment. In addition, the sink should be located in the anteroom to serve as a reminder to wash hands before entering or leaving the cubicle. Normally, water does not transmit nosocomial pathogens unless it is aerosolized. To that point, each cubicle should have a sink with faucets designed not to create aerosols and a toilet housed in a cabinet, so that after use it can be put back and flushed thereby limiting aerosolization. Covered hampers for soiled linen and waste should be kept in the cubicle itself, physically separated from the clean storage area of the anteroom.

Another aspect of the critical care environment is the ventilation system. Although the air in modern facilities contains minimal organisms as a result of centralized, filtered, nonrecirculating ventilation systems, there are instances when the number of spores in the air is significantly increased, such as during construction periods.

Adjustment in air changes, frequent cleaning, and the changing of air filters is a benefit to the critical care patient. Temperature and humitidy have a definite effect on the critically ill patient's evaporative water loss, wound healing, and the likelihood of pulmonary infection. The design of the unit should be such that the requirements set by the U.S. Hill-Burton Act (a temperature range of 21–23.5° C; a relative humidity between 30–60%) can be easily met (5).

Infection Control Program

An effective critical care infection control program is an integral component of quality patient care. The main goal of the program is to lower the risk of nosocomial infection for each patient in the unit. The priority for the program is to provide the safest possible environment for critical care patients, visitors, and staff.

Regardless of the size or the type of critical care unit, there are four components that should be present in an effective infection control program: surveillance, reporting, prevention, and control. The key person in the development and maintenance of an effective program is the infection control practitioner. The role of this practitioner is to be responsible for developing and carrying out the directions of the program.

Quality assurance is the important process that involves evaluating the critical care unit's infection control program to determine if standards are being met. It is a two-step process: (a) comparing standards with the actual results; and (b) making recommendations for improvements in the delivery of care. Critical care nurses possess the expertise necessary to establish outcome criteria for their patients. Quality assurance activities should be an integral part of nursing practice. It is the nurse manager's responsibility to promote an environment supporting the objectives and mechanisms for quality assurance to occur at the bedside.

Surveillance and Reporting

Surveillance in the critical care setting is a systematic, dynamic, continuing observation of the occurrence of infections and the conditions or events that have an impact on the risk of such occurrences. Reporting is the analysis and distribution of the gathered data to those persons who can use the information to take appropriate action.

It is not cost-effective to allocate personnel and financial resources for the collection of surveillance data that are not being used to prevent and control infections. Surveillance must be integrated into the unit's infection control program to assure its impact in the reduction of costly nosocomial infections. It is essential that the unit's surveillance program be derived from clearly stated outcome objectives. In other words, when evaluating the need for surveillance activities, it is necessary to ask how well this particular action reduces infections within the critical care unit.

In the achievement of the ultimate outcome, which is that of assuring quality care through the reduction of infections within the critical care environment, process standards must be identified and stated. These standards, if met, will reduce the risks of infection. Indicators related to these process standards may include any or all of the following: establishing baseline rates, identifying epidemics, influencing staff behavior or practices, evaluating interventions for effectiveness, meeting accrediting and governing standards, and participating in research.

The establishment of baseline rates provides knowledge of the endemic rate of nosocomial infections for the unit. Ongoing surveillance identifies deviations from the baseline rate. Alerted personnel should search for causes, thus preventing outbreaks.

Surveillance data can be used most effectively to convince the critical care staff that a problem exists and that recommended changes in practice should be adopted. As practice changes and becomes established, surveillance can be used to evaluate the effectiveness of the changes.

The Joint Commission on Accreditation of Health Care Organizations (JCAHO) directs hospitals to have ongoing surveillance systems to identify sources of infections, to discover epidemics, and to establish endemic levels for nosocomial infections. JCAHO outlines the following basic minimum infection control functions which are applicable to the role of nurse manager who is central in maintaining infection control standards:

- Define nosocomial infections for surveillance purposes;
- Determine the type of surveillance required;
- Monitor the surveillance results;
- Maintain reporting and communication of trends mechanism;
- Intiate corrective action;
- Evaluate and review infection control standards.

Surveillance data can validate research designed to decrease costs in the critical care unit without compromising patient care (e.g., extending the interval between central line intravenous tubing changes or respiratory breathing circuit changes). This is but one example of how surveillance can be used to support research projects designed to provide information that can ensure cost-effective, quality patient care.

It is important that critical care personnel utilize every possible measure in order to control infections. By monitoring all three factors in the chain of infection—the host, the agent, and the mode of transmission—this is possible. Infection control is an integral component of managing the environment in critical care nursing.

REFERENCES

1. Bennett J, Brachman P (eds). Hospital Infections, 2nd ed. Boston: Little, Brown, 1986.
2. Castle M. Hospital Infection Control. New York: John Wiley, 1980.
3. Wenzel R. Prevention and Control of Nosocomial Infections. Baltimore: Williams & Wilkins, 1987.
4. Montgomery B, Luce J. Infection monitoring. Respir Care 1985;30:489–499.
5. duMoulin G. Minimizing the potential for nosocomial pneumonia; architectural, engineering, and environmental considerations for the intensive care unit. Europ J Clin Microbiol Infect Dis 1989;8:69–71.

Chapter 6
Establishing a Clinical Nursing Research Program

GLADYS M. CAMPBELL, MARIANNE CHULAY

The expenditures of resources for a clinical nursing research program has long-term payoffs that include recruitment and retention, professional image enhancement, positive impact in the practice setting, and staff development. This chapter provides an overview of some of the benefits of a clinical research program and reviews the administrative supports and resources necessary for a research program, along with describing the climate necessary at the practice level before beginning the research process. The emphasis is on staff nurse involvement in a unit-based approach to clinical research. The greatest rewards from a research program are reaped when staff nurses have an integral role in the program.

BENEFITS OF CLINICAL RESEARCH

Recruitment and Retention

A clinical nursing research program can be viewed by nursing departments as an adjunct to recruitment and retention efforts. Financial pressures cause many departments to eliminate programs that do not have an immediate impact on care delivery. Any savings generated from this elimination are then used to increase staff salaries. The hope is that increased salaries are immediate retention incentives and enhance the department's ability to recruit competitively.

The recruitment and retention literature supports the notion that financial incentives do not alone improve either retention or overall job satisfaction [1,2]. Although structural issues like salary and scheduling options have a very real impact on nurses and the profession, job satisfaction and the potential for retention are also linked to issues of professionalism [2,3]. Issues like autonomy, responsibility, and potential for professional advancement have been shown to affect retention of nurses significantly [3]. Friss [3] has identified that nurses enter employment with expectations of advancing their knowledge and skills, being challenged, and having the opportunity to advance and to be recognized [4]. Turnover within nursing may be related to the fact that some of these expectations are unmet.

Active staff nurse involvement in a clinical nursing research program is one possible way to meet these expectations. Involvement in research allows staff nurses to use and advance their knowledge and skills and to provide an opportunity for achievement and recognition. This opportunity provides a new, challenging dimension to the job. In addition, retention can be positively affected by a clinical research program.

Professional Image Enhancement

It is not uncommon to see positive changes in staff nurses who are doing research. Nurses involved in research begin to change their view of themselves, from "just a nurse" to a nurse making a valuable contribution to the body of nursing knowledge. That change in self-perception becomes contagious as more and more staff members within a unit or department begin to participate in research. The work environment transforms in a positive, professional direction.

As nurses present research findings at conferences, poster sessions, or workshops, they not only have the opportunity to be recognized for their work, but to display a positive professional image to those outside the institution. In this way, their image becomes the image of the nursing department. Nurses from other organizations will be attracted to this positive image. Nurses within the department have increased pride in their continued association with a progressive nursing service that is recognized by other organizations. In this way a nursing research program can enhance professional image.

Positive Impact in the Practice Setting

The value of nursing can be measured by relating patient care outcomes to the quantity and appropriateness of nursing care delivered (5). This measurement of nursing intervention against patient outcomes is valid only when nursing interventions are grounded in scientific research. How can nurses as a profession define their unique contribution to patient well-being if the relationship between what nurses do and patient response cannot be proven? Thus, clinical research is an indispensable component of professional nursing practice and requires the commitment of nurses practicing at every level and in every setting. Oberst (5) found that "5–10% of a profession's total effort must be devoted to research if progress is to be made. At the present time, the number of nurses engaged in research still falls far below the level. Both validation of nursing practice and the advancement of the state of nursing knowledge will require that increasing numbers of nurses become actively involved in the study of nursing problems."

The American Nurses Association has published *Guidelines for the Investigative Functions of Nurses* (6). These guidelines support investigative activities of nurses prepared in all programs and support nurses having an active role in the processes of developing scientific knowledge and incorporating that knowledge into practice. Table 6.1 describes the guidelines for investigative activities.

Nurses in clinical practice make decisions about nursing interventions many times a day, yet rarely have a rationale for decision-making beyond personal experience or intuition. Although experienced nurses generally have very good "intuition," this does not diminish the need for a scientific basis for practice. The profession can only be advanced if nursing practice can be validated through nursing research.

Research findings have impacted or changed many nursing interventions. In critical care, areas of practice such as temperature monitoring, patient positioning during hemodynamic monitoring, methods of endotracheal suctioning and timing of preoperative teaching, to name a few, have all been improved by nursing research. Nursing needs to continue to define standard nursing interventions and then test through research the ability of these interventions to achieve the expected patient outcomes.

In critical care, nurses are often overwhelmed with the level of technology common to the workplace. In critical care units, it often seems that a "more is better" philosophy exists in relation to technical equipment. Each year new technology is introduced with claims of improving the efficiency and effectiveness of care. Critical care nurses are in a position to test these claims. The appropriate evaluation of technology through research cannot only lead to significant cost savings and decreased labor intensity, but may actually prevent potential hazards to the patient. In critical care, many technical devices are currently utilized with minimal clinical research on their effectiveness. Devices such as pulse oximetry, bedside laboratory technology, and new temperature monitoring devices are often introduced with minimal testing in the critically ill patient population. Before critical care nurses use these types of equipment to make decisions on the appropriateness of nursing interventions, they need a high level of confidence that the information or intervention gained from this equipment is valid and reliable.

Staff Development

Staff development may be the underlying goal of a research program or an unexpected benefit. Through the mentorship of a nurse researcher, staff nurses can increase their knowledge and skills related to research proposal development, data collection, and analysis. They can increase their ability to recognize areas of needed research in the practice arena and to evaluate the significance of completed research. They are more inclined to challenge nursing "rituals" and to incorporate valid research findings into their practice. Staff members who have been directly involved in clinical research also have the opportunity to refine writing skills and to develop comfort with public speaking.

Table 6.1. Investigative Function of Nurses[a]

Guidelines for the Investigative Function of Nurses

Associate Degree in Nursing[b]
1. Demonstrates awareness of the value or relevance of research in nursing.
2. Assists in identifying problem areas in nursing practice.
3. Assists in collection of data within an established structured format.

Baccalaureate in Nursing[b]
1. Reads, interprets, and evaluates research for applicability to nursing practice.
2. Identifies nursing problems that need to be investigated and participates in the implementation of scientific studies.
3. Uses nursing practice as a means of gathering data for refining and extending practice.
4. Applies established findings of nursing and other health-related research to nursing practice.
5. Shares research findings with colleagues.

Master's Degree in Nursing
1. Analyzes and reformulates nursing practice problems so that scientific knowledge and scientific methods can be used to find solutions.
2. Enhances the quality and clinical relevance of nursing research by providing expertise in clinical problems and by providing knowledge about the way in which these clinical services are delivered.
3. Facilitates investigations of problems in clinical settings through such activities as contributing to a climate supportive of investigative activities, collaborating with others in investigations, and enhancing nursing's access to clients and data.
4. Conducts investigations for the purpose of monitoring the quality of the practice of nursing in a clinical setting.
5. Assists others to apply scientific knowledge in nursing practice.

Doctoral Degree in Nursing or a Related Discipline
  A. Graduate of a practice-oriented doctoral program
    1. Provides leadership for the integration of scientific knowledge with other sources of knowledge for the advancement of practice.
    2. Conducts investigations to evaluate the contribution of nursing activities to the well-being of clients.
    3. Develops methods to monitor the quality of the practice of nursing in a clinical setting and to evaluate the contributions of nursing activities to the well-being of clients.
  B. Graduate of a research-oriented doctoral program
    1. Develops theoretical explanations of phenomena relevant to nursing by empirical research and analytical processes.
    2. Uses analytical and empirical methods to discover ways to modify or extend existing scientific knowledge so that it is relevant to nursing.
    3. Develop methods for scientific inquiry of phenomena relevant to nursing.

[a] From the ANA. Guidelines for the investigative function of nurses. Kansas City: American Nurses Association, 1981.
[b] This language was developed as part of the work of the ANA Commission on Nursing Education and was included in the report of that commission to the 1980 ANA House of Delegates.

ADMINISTRATIVE READINESS TO CONDUCT RESEARCH

The establishment of a clinical research program requires the support of the nursing department at large (5,7). A departmental commitment means making available the financial, personnel, and equipment resources that are required for a research program. This potentially large commitment of resources during a time of fiscal constraint can seem unrealistic without a departmental "vision" of the future that uses a research program to help meet organizational goals. A free-standing research program that is not responsive to the goals of the department is potentially cost-ineffective, not clinically applicable, and beneficial only to those who are directly involved (8).

A research program can assist in accomplishing future-directed strategies that are efforts to pursue a specific vision or design for the future. Future-directed strategies may focus on areas like recruitment and retention, development of nursing care as a competitive edge for the organization, outcome-focused quality assurance, alternative models of care delivery, clinical practice problems, and development of a professional climate. Enhancing the image of the nursing department within the profession also enhances recruitment abilities. Visionary leadership in the nursing department may be the strongest asset in developing a research program.

Once the organization's goals have been successfully addressed, the nursing department can extend its vision to include the larger professional community. This vision may include contributing to the body of knowledge in nursing, collaborative efforts between education and service settings, improving the professional image of nursing, and/or establishing a scientific basis for practice (7). The movement of a nursing department beyond survival strategies and into broader professional issues may in and of itself improve the professional climate and pride within that department. Once a departmental commitment to research has been made, an administrative structure that supports research must be established. Table 6.2 lists indicators of administrative support for clinical research. A decentralized nursing department environment may provide a more supportive climate because decentralization encourages independence in staff and delegaton of decision-making to nursing staff, which, in turn, provides for self-motivation. Staff nurses working in a centralized nursing department may feel too inhibited by the static organizational structure to risk involvement in the research process.

Staff nurses are generally interested in research but many have neither the education nor the experience to pursue research independently. Allowing the unprepared but enthusiastic staff nurse to forge ahead risks producing only frustration and disenchantment with the research process. In addition, nurse managers are often ill prepared for research themselves, much less able to act as mentors to staff members in the research process. Even if nurse managers have such skills, rarely will they have the time to devote to research mentorship.

Mentoring of nursing staff in the research process most naturally falls to the clinical nurse specialist (CNS) (5,9). For the CNS to be successful in this role, several conditions need to exist. The CNS must have the support of the nurse executive and, in addition, particularly the support of the critical care nurse manager. The CNS must have expertise in the research process and must also have a desire to participate in research and a willingness to share expertise with others (10). The nurse executive who is establishing a research program needs to consider actively recruiting a pool of CNSs with research experience and doctoral preparation if possible. Ideally, the CNS should be based in the critical care unit rather than be accountable for multiple units. This enables the CNS to focus on the research needs of the patient population and to set research as a priority and not as a "frill" for which there is never time (5). In organizations where budgetary constraints do not allow for additional resources, the nurse executive may consider reorganizing and restructuring assignments to create a research position. In the decentralized environment this may be easier because nurses in leadership positions and staff nurse positions have been more autono-

Table 6.2. Indicators of Administrative Support for Clinical Research[a]

Inclusion of a statement supporting research in the philosophy
Presence of an active nursing research committee with a membership that is enthusiastic and knowledgeable about research
Identification of annual goals related to research
Appointment of personnel to the staff who conduct research as their primary responsibility
Dedication of resources to provide support services (e.g., consultants, secretarial support, photocopying, money to attend conferences, etc.)
Dedication of educational money to foster the enhancement of research skills of the staff
Inclusion of research-based performance competencies in position descriptions
Creation of a climate that fosters inquiry and creativity
Promotion of networking activities among nurses and other health care disciplines
Promotion of innovative change in nursing practice based on systematic inquiry

[a] From Lieske AM. Utilizing clinical research as a human resource management tool. In Lewis EM, Spicer JG eds. Human Resource Management Handbook: Contemporary Strategies for Nursing Managers. Rockville: Aspen Publishers, 1985: p 77.

mous and independent and are better prepared to take on different and new responsibilities.

RESOURCES

For staff members to participate in nursing research they must be supported with time, mentorship, clerical assistance, and administrative encouragement in order to achieve a successful outcome. Recognition of these requirements and the use of creative strategies to acquire additional resources enables the nurse manager to overcome obstacles in this area.

Personnel

Key tasks need to be accomplished during the development and implementation of the clinical research project. Secretarial support and word processing capabilities are needed to expedite the development of protocols, abstracts, and publications. If the investigators have not had prior experience in clinical research, it is necessary to solicit consultation from a nurse researcher.

The consultation of a statistician is required during the protocol development stage. The complexity of nursing research today almost dictates the early involvement of this individual in protocol development. In addition to providing guidance on the appropriate statistical tests, the statistician can assist with sample size projections and the development of data collection forms. Very few clinical facilities have statistical consultants on staff but local colleges and universities are excellent resources. Financial reimbursement of this individual may not be necessary if coauthorship rights can be negotiated.

Another personnel requirement in clinical research is for data collectors. Most nursing budgets do not include monies for research data collectors, so creativity is needed to meet this essential personnel requirement. Sources of data collectors include the researchers themselves, staff nurses, graduate nursing students, and temporary employees. For many studies it is possible to incorporate data collection requirements into the usual implementation of nursing interventions. For example, the collection of hemodynamic measurements for a research study can be coordinated with the usual routine for obtaining hemodynamic measurements. If the time required for data collection from a subject is of short duration, this coverage can be incorporated into shift assignments. Shift overlap times can also be successfully utilized to obtain data. Another source of data collectors is graduate nursing students. Negotiating with faculty teaching research courses to arrange a practicum in clinical research can be valuable to the critical care service setting as well as to the academic program.

Financial

Depending on the topic of study, clinical research can be accomplished with minimal financial support. Studies that examine current clinical practice situations often require minimal, if any, outlay of fiscal resources. For example, determining the accuracy of low volume injectates for cardiac output determinations could be done with existing equipment and supplies.

When financial support is required for clinical research, a major source is grants from professional, private, and governmental groups. In view of the limited fiscal resources of these agencies and the large number of experienced nurse researchers competing for the funds, it is unlikely that a neophyte researcher would obtain funding.

Several other resources for financial support are available to the beginning researcher besides research grants. Resource development techniques can be used once the research project is fairly well defined. For example, local businesses, community volunteer organizations (Jaycees, Lions Clubs, women's clubs), or hospital volunteer groups might be interested in contributing to the project. In addition, specific items or services that are required for the research could be requested from several groups, so that "funding" of the study would be a compilation of several sources.

Another source of support is "creative financing." This involves the use of borrowing or trading services with other departments in the hospital in order to finance the research. Negotiating with heads of departments to "trade" services might be a very effective way to fund a study. Continuing education classes for the respiratory therapy department might be exchanged for free arterial blood gas analysis for an endotracheal suctioning study. Another approach would be to collaborate with the respiratory therapy department to develop and implement research on endotracheal suctioning.

Miscellaneous

It is important to realize that a great deal of emotional support is required throughout the research process. Unexpected roadblocks do emerge, priorities of projects change, and the time requirements are usually greater than originally estimated. These inevitable occurrences can decrease the researchers' enthusiasm and ability to drive the project to completion. The leadership must recognize the importance of ongoing support for the research team members throughout the research process.

CLIMATE FOR CLINICAL RESEARCH

The implementation of a clinical research protocol requires the support of all the nursing staff of the unit. It is important to establish a sense of commitment and excitement for the research process and to communicate the importance of the staff's integral role in the conducting of clinical research. One major strategy to accomplish this is the active involvement of the nursing staff in the development of the research project from its inception (11). Discussions on the scientific basis for such interventions as chest physiotherapy for lobar atelectasis, or hyperoxygenation and hyperventilation during endotracheal suctioning, or the techniques used to obtain accurate cardiac output determinations may assist in highlighting the role that research plays in the provision of nursing care to critically ill patients.

Another way to set the climate for research is to have staff nurses who have actively participated in research projects discuss the experience and its benefits and limitations. In describing their involvement, motivations, and initial inexperience with the research process, a real and credible example of a successful outcome is presented. Even more important is the natural sharing of enthusiasm that occurs in this type of situation.

Commitment and enthusiasm can also be developed by actively involving staff members in the identification of the research question to be studied. Their perspective on the clinical practice problems that need to be addressed assists in the identification of important issues in clinical practice, as well as ensures their early involvement in the research process.

BEGINNING THE RESEARCH PROCESS

As a nursing department plans for a research program and begins putting structural supports in place, it must also recruit nurses who are prepared and positioned to set that program's goals into operation. These nurses need the educational credentials, experience, enthusiasm, and communication skills to mentor others in the research process. With recruitment completed and the structure in place, the research process is ready to begin.

The "research mentor" needs to decide who is going to be involved in the first research project for the program. Will individuals be allowed to self-select based on interest, or will they be intentionally selected based on position in the unit or department? Important characteristics or criteria for selecting members of the research project include enthusiasm, clinical competence, professional respect, maturity, and good verbal and written communication skills.

It may be advisable to have the first research group participants consist of nurse managers and CNSs. By building on their interest and enthusiasm, a core of leadership personnel can be developed to support future research endeavors and to be future "research mentors." The first activities of nurses interested in research should include: research-based grand rounds, journal clubs, development of a research library, distribution of abstracts, and facilitating educational courses on the research process. Once nurse managers and CNSs have had an opportunity to develop some knowledge and skill about the research process during these activities, they are in a better position to support staff nurse involvement in the research process actively.

After the participants are selected, they need to be made aware of the level of time commitment the project will require. The total number of months or years for project completion needs to be estimated and clearly communicated. In addition, the amount of time per week required for the various phases of the project needs to be considered. Once time requirements are projected, potential project participants need to assess their abilities to commit to the research project. Sometimes having individuals list and prioritize their current professional commitments helps them determine their abilities to participate in the research.

The purpose of an initial research project is to increase the knowledge and skills of the participants related to the research process. The project should address an important clinical need that can be seen as valuable and relevant to practice so that commitment and enthusiasm is easy to maintain throughout the life of the project. The important outcome of the first project should be to build confidence, commitment, and enthusiasm in a group of potential nurse researchers.

The selection of the research question is perhaps the most important aspect of the success of a clinical research project. An excellent source for listings of research questions thought to be important in critical care nursing are publications on research priorities (12) and reviews of research on nursing interventions (13–15). Table 6.3 lists research priorities for critical care nursing as identified by Lewandowski and Kositsky (12). Given the fiscal and personnel resources available to the team, a question should be chosen which is clinically important and easily studied. This substantially diminishes the potential roadblocks, improving the likelihood of successful completion of the project. The majority of clinical research projects that fail to be completed have an overambitious research question.

To ensure that there is an adequate source of available subjects, a research question which would apply to large groups of patients in the critical care unit should be selected. If the research question is directed at a patient population which, while interesting, is infrequently admitted to the critical care unit, study completion time will be greatly prolonged (16). Enthusiasm for research is easier to maintain if tangible results are apparent.

Another potentially viable method to determine research questions is the use of journal clubs. Initially, journal clubs can be used to provide a forum to discuss prior research efforts and directions for future research. These discussions, or critiques, help identify areas where methodological problems exist in previous studies and where additional research is needed.

Another technique is to identify research questions through the use of a focus group format. With this approach, a focus group facilitator provides questions that are directed at identifying the staff's perception of the nursing care problems in their patient population. For example, in discussing the nursing care problems of patients in a cardiac surgical intensive care unit, the staff might identify the high incidence of postoperative pulmonary complications. A discussion of the research basis for nursing interventions directed at these complications could follow, with some "challenges" to the scientific basis of these interventions. The next group session would then focus on previous research directed at prevention of pulmonary complication. In reviewing this literature, some of the current nursing practices will be found to have no scientific basis. This provides an excellent basis for a unit research question (17). In order to assist the group in focusing on a realistic and clinically relevant research question, the focus group facilitator must be current in the content area.

Table 6.3. Research Priorities for Critical Care Nursing[a]

Sleep patterns of critically ill patients
Burnout
Orientation programs
Effects of stimuli on intracranial pressure
Weaning patients from ventilators
Patient classification systems
Incentives to retain nurses
Staff stress
Patient positioning and its effects on patients' cardiovascular and pulmonary status
Staffing patterns
Preventing infections in patients with invasive lines and/or undergoing invasive procedures
Suctioning patients on ventilators
Research utilization
Methods of assessing and relieving pain

---

[a] From Lewandowski L, Kositsky A. Research priorities for critical care nursing: A study by the American Association of Critical-Care Nurses. Heart Lung 1983; 12:35–44.

Another consideration in this area should be the practicality of the research question. Are the necessary research tools/instruments to evaluate the research question available? The development of research instruments, while seemingly simple, are actually very complex tasks if valid and reliable tools are to be developed. For example, if an appropriate tool to measure psychosis in the critically ill patient cannot be found in the literature, neophyte researchers would be well advised to select a different research topic.

A very successful approach to "finding the right research question" is to consider replication studies (18). Here the neophyte researcher follows the recommendations for future research offered at the completion of the research reports. Studies can be replicated in the same patient population, a new population, or with improved or expanded methods to answer a research question more clearly and completely. For example, a few studies have investigated the effect of body positioning on the measurement of hemodynamic parameters in critically ill patients with normal pressures. Repeating these studies in patients with abnormal pulmonary artery pressures would broaden the generalization of findings. Replication research provides an opportunity for beginning researchers to enter the research arena more easily to study an important question with proven methods, as well as to contribute to the nursing profession's knowledge base.

PROGRAMS THAT MAY NOT SUCCEED

Many nursing departments have not succeeded in establishing nursing research programs. These programs have failed to produce a sufficient volume of quality research and have not stimulated the incorporation of research into practice. Thus, it is of some value to examine research programs that may have difficulty providing the research climate and desired outcomes.

One type of research program that can be problematic is if a nurse researcher builds a program around personal research interests, thereby not recognizing clinical research ideas generated by staff nurses. This type of program usually does not enhance achievement of nursing department goals and becomes a freestanding program apart from the department goals.

These programs fail for three main reasons. First, the research program is person-dependent. When the researcher leaves the institution, the program ceases to exist. Second, a research program that does not complement the goals of the department has trouble justifying its existence. Last, valuable research questions have to be generated by the staff nurse who can impact patient care outcomes. When staff nurses are left out of the research process, nurses may become disenchanted with a research program that does not incorporate and develop their expertise.

Another type of program that frequently does not succeed is the academically focused research program. This program usually has an academician as its department head, with graduate students as data collectors in the service setting. These research studies often have little immediate practical relevance. Hinshaw (19) discusses the differences between an academician's and a clinician's perspectives on research:

"Academicians are involved in the generation, construction, and testing of nursing theory—a body of knowledge whose ultimate reason for being is to guide practice. Clinicians are involved with actual practice problems, a major source for the generating of nursing theories. The clinician, however, primarily focuses on the need for an accurate data base from which to address an immediate, recurring problem. Clinicians are most concerned that data are relevant to and can be utilized in practice, whereas academicians are more concerned that data are generalized within the theoretical system under study" (19).

Nurses practicing at the bedside can develop negative attitudes about research when they perceive studies as being academic and clinically relevant. In addition, such studies are often hard to get through hospital-based research review boards because of their lack of clinical relevance.

FUTURE PERSPECTIVES

Research in nursing is still in its infancy. Although this may be frustrating to the profession, it creates multiple possibilities for the future. In the service setting, the most immediate need for research is in the area of actual clinical practice. Expected outcomes and nursing interventions associated with accepted nursing diagnoses need to have a basis grounded in research. The efficacy of frequently utilized nursing interventions need to be documented. Nursing "rituals" that are ineffective and time-consuming need to be eliminated. There is a tremendous need to replicate and validate research in diverse populations.

More nursing departments are adopting an outcome-focused approach to care. Descriptive research on diagnostic-related groupings (DRGs) would allow for analysis of patients that did not have the expected length of stay for the DRG. In this way, nursing research can be directly linked to departmental quality assurance.

Certainly there is a need for valid, applicable, relevant theories of nursing practice. Nursing academicians may be better able to establish a link from theory to practice through increased involvement in clinical settings. Joint appointments between the service setting and academic institution could prove beneficial to researchers in both areas.

The largest area of need in nursing research is to bridge the gap between research and practice. This cannot occur unless research is integrated into practice by the nurse at the bedside. The only way to assure that staff nurses take an interest in and accept research findings is to involve them actively in the research process.

REFERENCES

1. Everly G, Falcione R. Perceived dimensions of job satisfaction for staff registered nurses. Nurs Res 1976;25:346–348.
2. Weisman C. Recruit from within: Hospital nurse retention in the 1980's. J Nurs Adm 1982;12:24–31.
3. Friss L. Hospital nurse staffing: An urgent need for management reappraisal. Health Care Management Review 1982;4:21–27.
4. Seybolt J. Dealing with premature employee turnover. J Nurs Adm 1986;16:26–32.
5. Oberst M. Integrating research and clinical practice roles. Top Clin Nurs 1985;7:45–53.
6. American Nurses Association. Guidelines for the investigative function of nurses. Kansas City: American Nurses Association, 1981.
7. Marchette L. Developing a productive nursing research program in a clinical institution. J Nurs Adm 1985;15:25–30.
8. Fitzpatrick J. Room at the top . . . nurse administrators can direct the pace of research and clinical scholarship at their institution. Appl Nsg Res 1989;2:63.
9. Douglas S, Hill M, Cameron E. Clinical nurse specialist: A facilitator for clinical research. CNS 1989;3:12–15.
10. O'Brien M. Mentoring. In Cardin S, Ward C (eds). Personnel Management in Critical Care Nursing. Baltimore: Williams & Wilkins, 1989: p 107–123.
11. Hoare K, Earenfight J. Unit-based research in a service setting. J Nurs Adm 1986;16:35–39.
12. Lewandowski L, Kositsky A. Research priorities for critical care nursing: A study by the American Association of Critical-Care Nurses. Heart Lung 1983;12:35–44.
13. Riegel B, Forshee T. A review and critique of the literature on preoxygenation for endotracheal suctioning. Heart Lung 1985;14:507–518.
14. Barnes C, Kirchoff K. Minimizing hypoxemia due to endotracheal suctioning: A review of the literature. Heart Lung 1986;15:164–176.
15. Kiriloff L, Owens G, Rogers R, Mazzocco M. Does chest physiotherapy work? Chest 1985;88:436–444.
16. Sayner N. Research in the clinical setting: Potential barriers to implementation. J Neurosurg Nurs 1984;16:270–281.
17. Chulay M, White T. Nursing research: Instituting changes in clinical practice. Crit Care Nurs 1989;9:106–113.
18. Connelly C. Replication research in nursing. Int J Nurs Stud 1986;23:71–77.
19. Hinshaw AS. Collaborative nursing research. In Marriner A (ed). Contemporary Nursing Management Issues and Practice. St. Louis: CV Mosby, 1982: p 367–382.

Part IV

SUPPORT SYSTEMS SPHERE

Chapter 7

Ancillary and Support Services

SUSAN G. OSGUTHORPE

Critical care is optimally practiced when the nurse, physician, other health care providers, and ancillary and support services work together (1). All health care professionals have a responsibility to observe the patient, monitor the patient's progress, collect and evaluate physiological and laboratory data, and discuss problems and arrive at a plan of action (1). All providers have an area of expertise and skills that complement each other and, when all providers work together as a synergistic unit, the critical care provided is more powerful than any health care professional or service working alone (1). It is because of the vast amount of knowledge required and the complexity of the patient care encountered that no one person can observe and focus on everything at once nor can one individual master all the knowledge necessary (1). Therefore, a collaborative approach in critical care is indicated for optimal effectiveness.

Collaboration in health care has received a great deal of attention over the years and has come to mean different things to providers. Styles (2) addresses this phenomenon in the Styles Stipulation—"As a word gains in popularity it loses in clarity." Working together is the exact English translation of the Latin derivative of "collaboration;" this provides a useful working definition (2). The hierarchy of elements in collaboration (Fig. 7.1) described by Styles (2) includes people, purposes, principles, and structure. At the base of the hierarchy are the individual health care providers who share a common interest in health care and possess a fundamental compatibility (2). They have in common the specific purpose of providing effective and efficient health care to the critically ill patient (2). Established principles or ground rules relative to values, expectations, and communication operate within the structure of roles and relationships in the health care system (2).

In putting the collaborative elements into operation, Styles (2) describes stages or degrees of unity (Fig. 7.2). At the one extreme, no relationship exists and therefore no collaborative activity is seen. However, as the degrees of unity improve, interactive communication progresses to consultation and finally to seeking mutual consent before actions become unified policies and structures integral to the delivery of care, standards, policies, role functions, and accountability (2). In this context, collaboration lends itself to the managed care model described by Bower (3) as unit-based care that is organized to achieve specific patient outcomes within fiscally responsible timeframes (length of stay) while utilizing resources that are appropriate in amount and sequence to the specific case type and to the individual patient. The goals of managed care include:

1. To facilitate the achievement of expected and/or standardized patient outcomes;
2. To facilitate early discharge and/or reduced utilization within an appropriate length of stay;
3. To promote appropriate and/or reduced utilization of resources;
4. To promote collaborative practice, coordination of care, and continuity to care,
5. To promote professional development and satisfaction of hospital-based registered nurses; and
6. To direct the contributions of all care providers toward the achievement of patient outcomes (3).

ANCILLARY AND SUPPORT SERVICES

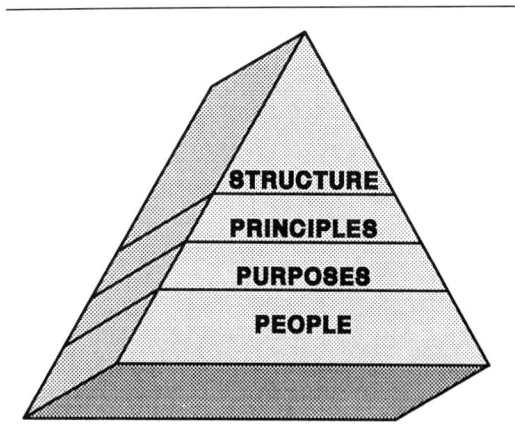

Figure 7.1. Hierarchy of elements in collaboration. (From Styles MM. Reflections on collaboration and unification. Image 1984;26(1):22.)

The managed care model will be utilized as a model to focus health care providers, support services, and therapeutic and diagnostic options to ensure quality care with a reasonable level of both benefit and cost (3).

McClure and Nelson (4) have identified the caregiver and integrator role functions of the nurse. They describe the caregiver role in terms of nursing responsibility in meeting patient needs in the areas of dependency, comfort, monitoring, therapy, and education. In meeting these varied needs, many health care providers in the specialties of pharmacy, occupational therapy, physical therapy, respiratory therapy, nutritional therapy, and social work must come together in an integrated manner to maximize the effectiveness and efficiency of the care provided. The critical care nurse must function in the integrator role to optimize the utilization of other health care providers and services in the care provided. Figure 7.3 illustrates the actual and potential of care provided with optimum utilization of all health care resources.

Ancillary health care provider functions are described in terms of pharmacy, respiratory therapy, physical therapy, occupational therapy, nutritional therapy, and social work services. In identifying optimal criteria for effective consultation and collaboration with these health care providers, the critical care nurse can more effectively integrate these specialized disciplines. Support functions are defined as the areas of facilities management, biomedical services, environmental services, laundry, and information systems. Last, critical care patients require extensive diagnostic services to facilitate the diagnosis and treatment of their illness. The diagnostic services such as noninvasive testing, radiology testing, or laboratory testing and their impact on the critical care patient and environment are discussed.

PHARMACY SERVICES

Clinical Role of the Pharmacist

The fundamental purpose of the profession of pharmacy is to serve society by ensuring the safe and appropriate use of drugs (5). It is significant that drugs are currently the most widely used therapeutic agents in the treatment of disease (6). Because drug therapy is a key component in determining patient care outcomes, the impact of ineffective or inefficient drug use will have consequences beyond the cost of acquiring the product (6). The pharmacist is accountable for promoting optimal use of drugs (including prevention of improper or uncontrolled use of drugs), and for providing authoritative drug information to other health care professionals, patients, and the public (5).

In ensuring optimal clinical outcomes from all drug therapy, the informational, clinical, and drug distribution components of a comprehensive pharmaceutical service are man-

A + B ⇒ A--------B ⇒ A———B ⇒ AB
No   Communication/   Consent/   Unification
Relationship Consultation   United Policy

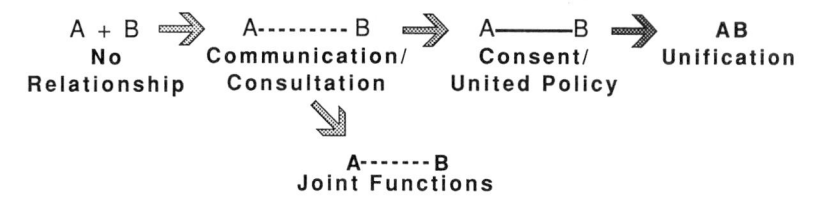

A-------B
Joint Functions

Figure 7.2. Conceptual models along the unity continuum. (Modified from Styles MM. Reflections on collaboration and unification. Image 1984;24(1):22.)

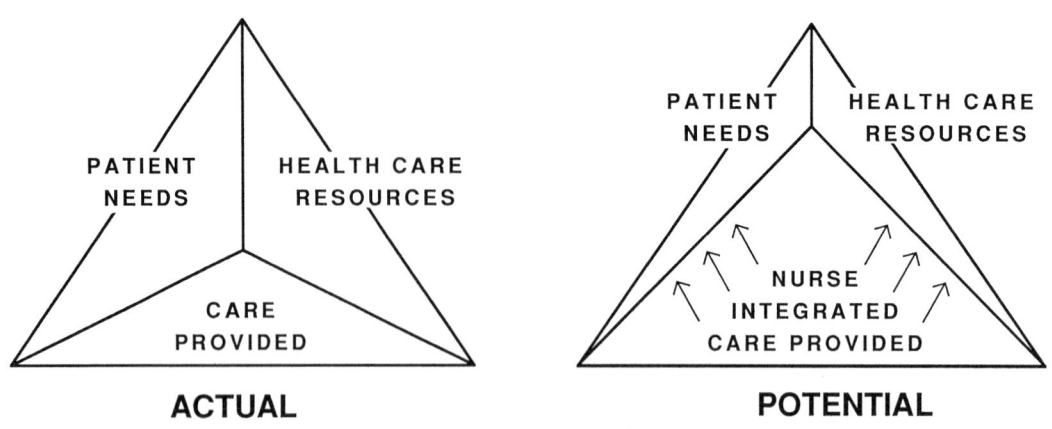

Figure 7.3. Unintegrated health care (*left*) and integrated health care (*right*). (Modified from Williamson JW, Alexander M, Miller GE. Priorities in patient-care research and continuing medical education. JAMA 1968;204: 93–98.)

aged as an integrated system. The American Society of Hospital Pharmacists (5) has identified an operational scope of clinical pharmacy services (Table 7.1).

The critical care nurse manager can utilize these examples of clinical pharmacy services to evaluate current and needed unit-based applications. Many of these clinical pharmacy services are particularly relevant guidelines for the critical care nurse manager and pharmacist to use as consultation considerations for the critical care staff nurse (Table 7.2).

Drug Distribution and Administration

The method of drug distribution is the essential element of pharmacy services. Without an effective, efficient, safe, and responsive system, control of the drug distribution and administration process is lost (6). Although many drug distribution systems are available, the most frequently encountered systems in critical care units include the unit-dose method and floor stock method.

In the unit-dose method, the pharmacists receive a direct copy of the physician's order, and medications are packaged in single-use containers and dispensed, whenever possible in a ready-to-administer form (6). A 24-hour supply of patient medications is dispensed in specially designed containers with individual patient drawers (6). The individual drawers are loaded into a cassette and sent to the critical care unit for storage in a medication area or mobile cart (6). The American Society of Hospital Pharmacists (7) has identified the following advantages of the unit-dose system over other methods of drug distribution:

- Reduction in the incidence of medication errors;
- Decrease in the total cost of medication-related activities;
- More efficient use of pharmacy and nursing personnel;
- Improved drug control and drug use monitoring;
- More accurate patient drug billing;
- Elimination or minimization of drug credits;
- Greater control by pharmacist over pharmacy workload patterns and staff scheduling;
- Reduction in unit drug inventories;
- Greater adaptability to computerized and automated procedures.

Intravenous (IV) admixtures (adding drugs to bulk containers of IV fluids) have been a labor-intensive procedure for critical care nurses in the past; however, the pharmaceutical industry has increasingly provided drugs intended for IV administration in a ready-to-administer dosage form (6). These IV admixture systems are very cost-effective in preparation time and money and complement the unit-dose system (6). Critical care managers should seek to involve the pharmacist in the development of policies and procedures for drug administration and IV therapy, nutri-

ANCILLARY AND SUPPORT SERVICES   65

Table 7.1. Clinical Pharmacy Services[a]

1. Drug therapy monitoring and communicating relevant findings and recommendations to other health care professionals who are also responsible for the patient's care. Drug therapy monitoring includes an assessment of:
   a. The therapeutic appropriateness of the patient's drug regimen;
   b. Therapeutic duplication in the patient's drug regimen;
   c. The appropriateness of the route and method of administration;
   d. The degree of patient compliance with the prescribed drug regimen;
   e. Drug-drug, drug-food, drug-laboratory, or drug-disease interactions;
   f. Clinical and pharmacokinetic laboratory data to evaluate the efficacy of drug therapy and to anticipate side effects, toxicity, or adverse effects; and
   g. Physical signs and clinical symptoms relevant to the patient's drug therapy
2. Documentation of pharmaceutical care in patient's medical record
3. Preparation of medication histories for the patient's permanent medical record or other data bases (e.g., medication profiles), or both
4. Provision of oral and written consultations with other health care professionals regarding drug therapy selection and management
5. Patient education and counseling regarding drug therapy and drug-related disease prevention
6. Participation in the drug therapy management of medical emergencies
7. Development of patient-specific drug therapy management plans and therapy endpoints
8. Control of medication administration in the patient-care area
9. Monitoring, detecting, documenting, reporting, and managing adverse drug reactions
10. Education of health care practitioners regarding drug use
11. Participation in drug-use evaluation and other quality assurance programs
12. Participative decision-making in pharmacy and therapeutics committee activities
13. Serving as a member of institutional review board, infection control, patient-care, drug-use evaluation, and other committees where input concerning drug use and drug policy development is required
14. Provision of accurate and comprehensive information about drugs and patient-specific drug information to other pharmacists, other health care professionals, and patients as appropriate
15. Initiation of and participation in drug and drug-related (e.g., medication administration devices) research, including formal clinical drug investigations

[a] Modified from ASHP *Statement on the Pharmacist's Clinical Role in Organized Healthcare Settings,* ASHP, (approved by the ASHP Board of Directors, November 16, 1988).

tional support services, cancer chemotherapy services, education and training of nursing personnel, infection and complication control, and drug or IV product selection and evaluation (6).

Despite widespread adoption of the unit-dose drug distribution system, most critical care units still require a labor-intensive floor stock system to minimize the delay of stat or first time administration of essential drugs.

Both the unit-dose and IV admixture services are usually provided as a centralized pharmacy service. In order to maximize the role of the clinical pharmacist and achieve drug use control, however, the pharmacist must be accessible to the patient and other health care providers in the areas where patient care decisions are being made (6). When pharmacists are available on the critical care unit in a decentralized pharmacy role, they may be (*a*) responsible for all clinical and drug distributions, (*b*) part of a decentralized unit-dose system using a pharmacy cart instead of the central pharmacy to restock unit-dose cassettes, or (*c*) managing a small satellite pharmacy in the critical care area (12).

It is common for critical care units to experience significant revenue loss related to medications that were not appropriately charged to

Table 7.2. Pharmacist Consultation Guidelines

1. Assessment of therapeutic drug regimen for individual critical care patients related to efficacy, interactions, reactions, and physical signs and symptoms related to drug therapy
2. Patient education and counseling related to drug therapy and drug-related disease prevention
3. Education of health care providers related to drug therapy

the patient because of oversight or a low priority on the "need to do" list when the critical care nurse is extremely busy with direct patient care priorities. The unit-dose system in conjunction with a decentralized unit-based pharmacist is extremely effective in managing drug cost control and utilization as well as providing more direct patient care time for the critical care nurse. Utilization of unit-based pharmacists greatly enhances the timeliness of processing physician orders (particularly related to stat and first doses of medications), formal and informal pharmacist consultation by other health care professionals, provision of patient and health care provider education, and problem resolution related to drug availability and administration. As a part of the health care team directly accountable for the care of the critical care patient population, the pharmacist becomes more familiar with each patient, the therapeutic treatment goals, and complicating issues related to drug administration (e.g., fluid restriction, limited IV access, poor renal or liver function, multiple/complex drug treatment protocols, or interaction/reactions). Although pharmacists and nurses generally concur that the pharmacy is primarily responsible for the drug distribution process and that nurses are primarily responsible for drug administration, Thompson (9) reported that pharmacists and nurses differ greatly about which professional holds primary responsibility for therapeutic drug monitoring and adverse drug reaction surveillance. The disagreement is related to unclear role clarification in this important area of collaborative practice. It is essential for the critical care nurse manager to facilitate role clarification between the pharmacist and the critical care nurse in each individual hospital setting to promote mutually beneficial roles in the therapeutic drug treatment process to enhance patient care outcomes (9).

In summary, a pharmacy service that utilizes a distribution system incorporating unit-dose and decentralized pharmacy concepts, as well as decentralized pharmacists to enhance clinical services, will be in a position to establish effective drug control and to have a positive impact on all elements of the drug use process (6).

RESPIRATORY THERAPY

Clinical Role of the Respiratory Therapist

Most patients in critical care units have some degree of respiratory difficulty, and respiratory problems are likely to occur while patients are in critical care units (10). The supine position decreases the functional residual capacity of the patient's lungs and contributes to the development of atelectasis, even if no other respiratory difficulty occurs (10). The primary function of the respiratory system is to provide oxygen for metabolism and to remove carbon dioxide, a waste product of metabolism (10). Secondary functions of the respiratory system include acid-base balance, speech, the expression of emotion through laughing, crying, and sighing, and maintenance of body water and heat balance (10).

Significant alteration or impairment of these critical functions requires expensive equipment, extensive monitoring of cardiovascular and pulmonary functions, and timely therapeutic interventions by a team of knowledgeable physicians, nurses, respiratory therapists, and other health care providers (11). In some patients, the respiratory problem is the primary illness, and in others, it is secondary to another illness (11). The critical care manager should facilitate role clarification by identifying both basic and complementary responsibilities of the physician, the critical care nurse, and the respiratory therapist on the critical care team. The role of the critical care nurse relative to respiratory therapy depends a great deal on the availability of respiratory therapy colleagues (11).

An operational scope of critical respiratory care interventions provided by respiratory therapy services is delineated in Table 7.3 (12).

Critical care nurse managers can utilize these examples of critical respiratory therapy interventions to evaluate current and possible unit-based applications in their own settings. Many of these respiratory therapy interventions are particularly relevant guidelines for the critical care nurse manager and respiratory therapist to use as consultation considerations for the critical care staff nurse (Table 7.4).

Table 7.3. Critical Respiratory Therapy Interventions[a]

1. Oxygenation Status: Assess the need for, recommend, and/or perform appropriate interventions which include: low flow oxygen therapy, high flow oxygen therapy, CPAP, mechanical ventilation, hyperbaric therapy, diuretic therapy, and treatment of pulmonary barotrauma
2. Bronchopulmonary Hygiene Status: Recommend, and/or perform the following interventions: airway suctioning, bronchial drainage, bronchial lavage, percussion, vibration, and humidity therapy
3. Pulmonary Function Status: Recommend and/or perform the following pharmacological interventions: inhaled sympathomemetic amines, anticholinergic drug administration, inhaled steroid therapy, antiasthma drug (cromolyn sodium) therapy, mucolytic agent administration, pentamidine aerosol administration, and ribavirin administration (note: topical or endobronchial instillation of selected drugs may also be appropriate)
4. Ventilatory Function: Assess the need for, recommend, and/or perform mechanical ventilatory support including: IPPB, CPAP, mechanical ventilation, jet ventilation, development of weaning plan, and inter- and intrahospital patient transport with ventilatory support
5. Artificial Airway Placement: Assess the need for, recommend, and/or perform the following interventions: check artificial airway placement and reposition if indicated, and endotracheal or tracheostomy tube intubation and extubation
6. Effectiveness of Oxygen and/or Mechanical Ventilation Therapy: Recommend and/or perform the following interventions: VD/VT studies, oxygen consumption and carbon dioxide production studies, intracranial pressure-carbon dioxide response studies, shunt studies, saturated venous oxygen monitoring, thoracic (lung) compliance-airway resistance evaluation, transcutaneous oxygen-carbon dioxide monitoring, pulse oximetry, exhaled carbon dioxide monitoring (capnography), positive end-expiratory pressure (PEEP) studies, arterial and venous blood gas analysis, hemodynamic monitoring, and weaning parameters
7. Cardiopulmonary Rehabilitation: Assess the need for, recommend, and/or perform the following interventions: rest and exercise blood gas analysis, rest and exercise oximetry, pulmonary function testing, and 7-minute walk test
8. Cardiopulmonary Resuscitation: Assess the need for, recommend, and/or perform basic cardiac life support and advanced cardiac life support
9. Special Procedures: Assist the physician, the critical care nurse, and other health care providers with special procedures which include, but are not limited to, bronchoscopy, tracheostomy, hemodynamic line insertion and/or monitoring, chest tube insertion, and sleep study testing

[a] Personal correspondence from James M. Smoker, RRT, Cochairman of the American Association of Respiratory Care Standards Committee, November 14, 1989. The above Critical Care Therapy Interventions are guidelines prepared for this publication by James M. Smoker, RRT and associates. The guidelines should not be interpreted as an official statement or document of the American Association of Respiratory Care.

Administration of Respiratory Therapy Services

The Joint Commission on Accreditation of Healthcare Organizations (JCAHO) (8) has defined standards for respiratory care services which include:

Standard 1. Respiratory care services that meet the needs of patients as determined by the medical staff are available at all times, are well organized, properly directed, and appropriately integrated with other units and departments of the hospital, and are staffed in a manner commensurate with the scope of services offered.

Standard 2. Personnel are prepared for their responsibilities in the provision of respiratory care services through appropriate training and education programs.

Standard 3. Respiratory care services are guided by written policies and procedures.

Standard 4. The respiratory care department/service has equipment and facilities to ensure the safe, effective, and timely provision of respiratory care services to patients.

Standard 5. Respiratory care services are provided to patients in accordance with a written prescription by the physician responsible for the patient and are documented in the patient's medical record.

Table 7.4. Respiratory Therapist Consultation Guidelines[a]

1. Assess the need for, recommend, and/or perform appropriate interventions related to:
   a. Oxygenation status
   b. Bronchopulmonary hygiene status
   c. Pulmonary function status
   d. Ventilatory function status
2. Assess the need for, recommend, and/or perform appropriate interventions related to artificial airway placement and support
3. Recommend and/or perform interventions necessary to assess the effectiveness of oxygen and/or mechanical ventilation therapy
4. Assess the need for, recommend, and/or perform interventions to assist the critical care patient with cardiopulmonary rehabilitation
5. Provide and participate in teams or committees for basic cardiac life support and/or advanced cardiac life support
6. Provide and participate in special procedures relative to the support of the cardiopulmonary system

[a] Derived from information contained in personal correspondence from James M. Smoker, RRT, Cochairman of the American Association of Respiratory Care Standards Committee, November 14, 1989. The consultation guidelines are not an official statement or document of the American Association of Respiratory Care.

Standard 6. As part of the hospital's quality assurance program, the quality and appropriateness of patient care provided by the respiratory care department/service are monitored and evaluated.

It is imperative that the critical care nurse manager, the critical care unit medical director, the respiratory therapy services manager, and the respiratory therapy service medical director regularly evaluate and implement recommendations relative to the therapeutic respiratory interventions. Collaborative management of this important aspect of critical care enhances the effectiveness of the care provided.

PHYSICAL REHABILITATION SERVICES

Many programs are available to promote the restoration of functional abilities of individuals with physical, cognitive, and sensoriperceptual impairment (13). Two key ancillary programs for critical care patients are the physical therapy service and the occupational therapy service. These ancillary services may be provided as an individual service or in conjunction with a comprehensive physical rehabilitation program (13). Regardless of organizational structure, the JCAHO (13) has required that organized physical rehabilitation services be available and should be based on an assessment of patient needs, should be provided by competent professionals, and should be delivered in accordance with a written plan for treatment. Because these two services are frequently managed under the auspices of physical medicine, it is important to clarify the services provided by physical therapy and occupational therapy to ensure appropriate consultation.

Physical Therapy Service

Physical therapy services provide identification, prevention, remediation, and rehabilitation of acute or prolonged physical dysfunction or pain, with emphasis on movement dysfunction (13). Physical therapy encompasses examination and analysis of patients and the therapeutic application of physical and chemical agents, exercise, and other procedures to maximize functional independence (Table 7.5) (13).

Key physical therapy services for the critical care nurse manager and physical therapy service to establish as consultation considerations for the critical care nurse appear in Table 7.6.

Frequently, nurses are aware of critical care patients with actual or potential physical disabilities caused either by the disease process or as a secondary factor in long-term treatment and support in the critical care unit. It is important for the critical care nurse to ensure patient evaluation and intervention by the physical therapy service. Multidisciplinary patient care conferences must occur with regularity in the critical care environment for optimal patient care to occur.

Occupational Therapy Service

Occupational therapy services provide for goal-directed purposeful activity to aid in the development of adaptive skills and performance capacities by individuals of all ages who have physical disabilities and related psychological impairment(s) (14). Occupa-

Table 7.5. Physical Therapy Services[a]

1. Physical therapy evaluation and assessment of the critical care patient before the provision of services
2. The determination and development of treatment goals and plans in accordance with the diagnosis and prognosis, with a treatment program aimed at preventing or reducing disability or pain and restoring lost function
3. Therapeutic interventions that focus on posture, locomotion, strength, endurance, cardiopulmonary function, balance coordination, joint mobility, flexibility, pain, and functional abilities in daily living skills
4. The application of modalities that include, but need not be limited to, heat, cold, light, air, water, sound, electricity, massage, mobilization, bronchopulmonary hygiene, and therapeutic exercise with or without assistive devices
5. Assessment and training in locomotion, including the use of orthotic, prosthetic, or assistive devices
6. Patient and family education
7. Monitoring the extent to which physical therapy services have met the therapeutic goals relative to the initial and subsequent examinations, as well as the degree to which improvement occurs relative to the identified physical dysfunction or the degree to which pain associated with movement is reduced

[a] From the Joint Commission 1990 Accreditation Manual for Hospitals. Chicago: Joint Commission of Accreditation of Healthcare Organizations, 1989: p 182–183.

Table 7.6. Physical Therapy Consultation Guidelines[a]

1. Evaluate and assess critical care patients for identification, prevention, remediation, and rehabilitation of acute or prolonged physical dysfunction or pain
2. Develop treatment goals and plan to prevent or reduce physical disability or pain and restore lost function.
3. Provide therapeutic interventions to improve locomotion, strength, endurance, cardiopulmonary function, balance, coordination, joint mobility, flexibility, pain control, and functional abilities in daily living skills
4. Provide patient and family education

[a] From the Joint Commission 1990 Accreditation Manual for Hospitals. Chicago: Joint Commission on Accreditation of Healthcare Organizations, 1989: p 182.

tional therapy is designed to maximize independence, prevent further disability, and maintain health (14). Comprehensive occupational therapy services are delineated in Table 7.7.

Critical care managers can utilize these examples of occupational therapy services to evaluate current and possible unit-based applications in their own settings. Occupational therapy services most appropriate for consultation in the critical care unit are presented in Table 7.8. The long-term patient outcomes can be significantly affected by early and appropriate consultation with occupational therapy services.

NUTRITIONAL SUPPORT SERVICES

Nutritional support of the critically ill patient has received increasing attention since 1974, when Butterworth chronicled the effects of malnutrition on hospitalized patients (15,16). Malnutrition is often a concomitant process in the critically ill patient resulting in pneumonia, wound dehiscence, an increased rate of sepsis, failure to wean from ventilators, and increased mortality (15).

Nutritional requirements for critically ill patients vary according to the disease process, the degree of injury, and the resulting hypermetabolism seen in the stress response to illness (15). Appropriate nutritional care of the critically ill patient must be initiated, monitored, and assessed by ongoing evaluation of nutritional status by the critical care nurse and the nutritional support team (15). The presence of the nutritional risk factors identified in Table 7.9 indicate the potential for malnutrition in a critically ill patient. The critical care nurse should consider obtaining nutritional consultation from the dietitian or the nutritional support service whenever one or more of these risk factors is present in the critically ill patient (15).

Campbell (15) prioritized metabolic support goals for critically ill patients (Table 7.10).

JCAHO (17) has defined standards for dietetic or nutritional support services. These include:

Standard 1. The dietetic department/service is organized, directed, staffed, and integrated with other units and departments/services of the hospital in a manner designed to ensure the provision of optimal

Table 7.7. Occupational Therapy Services

1. Occupational therapy assessment and treatment of occupational performance, including independent living skills, prevocational/work adjustment, educational skills, play/leisure abilities, and social skills.
2. The assessment and treatment of performance components, including neuromuscular, sensori-integrative, cognitive, and psychosocial skills
3. Therapeutic interventions, adaptations, and prevention.
4. Individualized evaluations of past and current performance, based on observations of individual and group tasks, standardized tests, record review, interviews, and/or activity histories
5. Achievement of treatment goals through the use of selected modalities and techniques including:
   (a) Task-oriented activities;
   (b) Prevocational activities; and
   (c) Sensorimotor activities
6. Achievement of treatment goals through:
   (a) Design, fabrication, and application of orthotic devices;
   (b) Guidance in the use of adaptive equipment and prosthetic devices;
   (c) Adaptation of the physical and social environment and the use of a therapeutic milieu;
   (d) Joint protection/body mechanics; and
   (e) Positioning
7. Patient and family education and counseling
8. Monitoring the extent to which occupational therapy services have met the therapeutic goals relative to assessing and increasing the patient's functional abilities in daily living and relative to preventing further disability

[a] From the Joint Commission 1990 Accreditation Manual for Hospitals. Chicago: Joint Commission of Accreditation of Healthcare Organizations, 1989: p 181–182.

nutritional care and quality food service.

Standard 2. Dietetic services personnel are prepared to conduct their assigned responsibilities through appropriate orientation, education, and training.

Standard 3. Written policies and procedures specify the provision of dietetic services.

Standard 4. The dietetic department/service is designed and equipped to facilitate the safe, sanitary, and timely provision of food service to meet the nutritional needs of patients.

Standard 5. Dietetic services are provided in accordance with a written order by the individual responsible for the patient, and appropriate dietetic information is recorded in the patient's medical record.

Standard 6. Appropriate quality control mechanisms are established (17).

Nutritional support service consultation should be obtained in the critically ill patient for:

1. Nutritional evaluation and assessment of the critically ill patient for nutritional risk factors and protein and energy status, and is accomplished through the use of anthropometric measurements, laboratory data, patient history, and clinical data;
2. Provision of oral nutritional support including special dietary considerations or restrictions;
3. Provision of nutritional support by enteral tube feeding;

Table 7.8. Occupational Therapy Consultation Guidelines[a]

1. Evaluate and assess the critical care patient with physical disabilities and related psychological impairment(s) relative to the development of adaptive skills and performance capacities
2. Develop treatment goals and plans to maximize independence, prevent further disability, and maintain health
3. Provide occupational performance therapeutic interventions
4. Provide patient and family education and counseling

[a] From the Joint Commission 1990 Accreditation Manual for Hospitals. Chicago: Joint Commission on Accreditation for Healthcare Organizations, 1989: p 181–182.

Table 7.9. Nutritional Risk Factors for Malnutrition

1. General
    Presence of conditions such as malabsorption syndromes, draining abcesses or wounds, protracted diarrhea, or vomiting which result in nutrient loss
    Presence of conditions such as fever, thermal injury, trauma, surgery, sepsis, chemotherapy, radiation therapy which increase the need for nutrients
    NPO
    Presence of food allergies, lactose intolerance, or limited food preference
    Patient that is more than 120 or less than 80% of ideal body weight, or experienced recent unexplained weight loss
    Patient on a modified diet such as clear or full liquid or diet restricted in sodium, calories, protein, and/or carbohydrates
    Patient receiving enteral or parenteral nutritional support
2. Gastrointestinal
    Patient with nausea, indigestion, vomiting, diarrhea, or constipation
    Patient with glossitis, stomatitis, or esophagitis
    Patient with mechanical difficulties chewing or swallowing
    Patient with a fistula
    Patient with a partial or total GI obstruction
    Patient that is edentulous or needs dental repair
3. Cardiovascular
    Patient with ascites or edema
    Patient with limited ability to perform activities of daily living
4. Genitourinary
    Patient whose input approximates output
    Patient with an ostomy
    Patient requiring hemo- or peritoneal dialysis
5. Respiratory
    Patient receiving ventilatory support
    Patient receiving oxygen via nasal prongs
6. Integument
    Patient with pressure areas on sacrum, hips, or ankles
    Patient with rashes or dermatitis
    Patient with dry or pale mucous membranes
7. Extremities
    Patient with pedal edema
    Cachexic patient as evidenced by decreased skin turgor, reduced buccal fat pads, or marasmic appearance

[a] Modified from Campbell SM. Nutrition. In Kinney MR, Packa DR, Dunbar SB (eds). AACN's Clinical Reference for Critical-Care Nursing, 2nd ed. New York: McGraw-Hill, 1988: p 339.

4. Provision of parenteral nutritional support;
5. Provision of patient and family education (15).

As the nutritional requirements of the critically ill patient vary greatly according to disease and clinical condition, it is important for the dietitian, nutritional support service consultant, critical care nurse, and other health care providers to evaluate carefully the nutritional needs of the patient (15). Once the nutritional therapeutic goals and plan are determined, careful monitoring and evaluation must continue to prevent complications arising from inadequate or overzealous provision of nutrients (15). Dietitians and nutritional support service consultants should evaluate all patients in the critical care unit on a regular, if not daily, basis. Collaborative efforts should ensure optimal nutritional therapy that maintains or improves the nutritional status of the critical care patient.

SOCIAL WORK SERVICES AND DISCHARGE PLANNING

Social work services provide for assessment and intervention relative to psychosocial factors and the social context in which the physically or emotionally disabled patient lives (18).

Table 7.10. Goals for Metabolic Support

---

1. Provide fluid resuscitation and support organ function; restoration of cardiovascular homeostasis in itself is useful in reducing catecholamine levels, thus decreasing the period of hypermetabolism which follows major tissue damage and acute illness
2. Repair wounds and damaged tissue; the degree of adrenergic response is related to the extent of the open wound; the hypermetabolic response evoked by catecholamines is abated when wounds are close, fractures immobolized, and septic and contaminated areas are drained
3. Provide nutritional substrates when the patient is stable and capable of using them for anabolism; this period does not begin until 24–48 hours after injury

---

[a] From Campbell SM. Nutrition. In: Kinney MR, Packa DR, Dunbar SB (eds). AACN's Clinical Reference for Critical-Care Nursing, 2nd ed. New York: McGraw-Hill, 1988: p 360.

Comprehensive social work services and discharge planning are delineated in Table 7.11 (19).

The social worker or discharge planner is charged with maintaining a patient-centered as well as problem-centered approach, patient advocacy, coordination of services within a facility, and a link between the facility and community (20).

Critical care nursing staff members should seek consultation from social services and discharge planning for the patient populations identified in Table 7.12 (21).

When it is anticipated that the patient will be going home from the hospital, the critical care nurse and social worker should assist the patient and family to plan to:

- Provide medications, treatments, and diet as prescribed by the physician for any of the patient's problems;
- Provide appropriate care and attention for bowel, bladder, and skin problems;
- Provide as much assistance as the patient needs with activities of daily living;
- Provide/obtain needed equipment and supplies; and
- Provide care consistent with the patient's

Table 7.11. Social Work Services and Discharge Planning

---

1. Assessment of the patient's personal coping history and current psychosocial adaptation to disability
2. Assessment of immediate and extended family members and other support persons relative to support networks
3. Assessment of housing, living arrangements, and stability and source of income relative to facilitating discharge plans
4. Implementation of intervention strategies designed to increase the effectiveness of coping, strengthen informal support systems, and facilitate continuity of care; these include:
    a. Discharge planning activities
    b. Casework counseling and therapy
    c. Group work focused on education and therapy
    d. Community service linkage and referrals
5. Monitoring the achievement of goals relative to discharge planning activities designed to meet basic sustenance, shelter, transportation, and comfort needs of patients and their families

---

[a] From the Joint Commission 1990 Accreditation Manual for Hospitals. Chicago: Joint Commission of Accreditation of Healthcare Organizations, 1989: p 186.

Table 7.12. Social Services and Discharge Planning Consultation Guidelines

---

1. Patients who are admitted from or may be transferred to a nursing home or specialty hospital
2. Patients in need of supportive follow-up treatment, teaching, and/or referral to community agencies
3. Patients followed by community agencies at the time of admission who may require continued follow-up after discharge
4. Patients with a history of repeated closely spaced readmissions
5. Patients with inadequate financial resources
6. Patients with age-related problems returning to their previous environment
7. Patients with congenital anomalies that impair/impede normal growth and body function
8. Patients suffering from multiple trauma
9. Patients with progressive neurological, neuromuscular, metabolic, cerebrovascular, pulmonary, or renal disease with resulting impairments
10. Patients with unexpected multiple births
11. Patients who have attempted suicide
12. Patients with problems related to substance abuse
13. Patients who are victims of child, spouse, or parent abuse

---

[a] Modified from Discharge Planning Update: An Interdisciplinary Perspective for Health Professionals. Chicago: American Hospital Association, 1980;1(1):24.

emotional or mentation problems if these are present (20).

If it is anticipated that the needed care cannot be provided at home, discharge planning with the patient and family will involve discussion of and planning for skilled nursing placement (20). The critical care nurse and the social worker need to provide a verbal and written report that includes the following information:

- The patient's present diagnosis with dates of treatment such as surgeries, chemotherapy, radiation therapy, or physical therapy;
- Secondary or additional diagnoses that require ongoing treatment;
- Assessment of the patient's level of functioning for completing activities of daily living;
- Other significant problems such as bowel or bladder incontinence, impaired skin condition, or visual and/or auditory handicaps (20).

To facilitate collaborative health-care discharge planning, Crittenden (20) suggests utilization of the format presented in Table 7.13.

The critical care nurse shares responsibility for discharge planning for critically ill patients with the social worker, the discharge planner, and other health care providers. It is particularly true of many critical care patients that "A discharge plan is more than a treatment program. . . . a discharge plan is a new way of life for many people" (22).

SUPPORT FUNCTIONS: FACILITIES MANAGEMENT

McLarney (23) defines facilities management as a comprehensive program designed to ensure a safe and comfortable environment for patient, staff, and visitors. Facilities management responsibilities include the functional areas identified in Table 7.14 (23).

Administration of Plant, Technology, and Safety Management

Although the engineering department assumes responsibility for many functional areas, environmental services, safety and security, and facilities planning are important services necessary for effective and efficient facilities management (23). Integration and implementation for many of these functional areas is articulated in the JCAHO (24) standards for plant, technology, and safety management, which state:

Standard 1. There is a safety management program that is designed to provide a physical environment free of hazards and to manage staff activities to reduce the risk of human injury.

Standard 2. There is a life safety management program designed to protect patients, personnel, visitors, and property from fire and the products of combustion and to provide for the safe use of buildings and grounds.

Standard 3. There is an equipment management program designed to assess and control the clinical and physical risks of fixed and portable equipment used for diagnosis, treatment, monitoring, and care of patients and of other fixed and portable electrically powered equipment.

Standard 4. There is a utilities management program designed to ensure the

Table 7.13. Collaborative Discharge Planning Format[a]

Patient	Medical Problem	Med/Nsg Needs	Social Needs	Psychological Needs
Mr. C	Angina	Meds, special diets, progressive exercise program, follow-up appointment	Less demanding workload, financial assistance	Help with fear of heart attack, stress reduction techniques

[a] From Crittenden FJ. Working with the staff alliance. In: Discharge Planning for Health Care Facilities. Los Angeles: Allied Health Publications at the University of California Extension, 1983: p 39–70.

Table 7.14. Facilities Management Functional Areas

---

1. Facilities engineering
2. Construction management
3. Building and equipment maintenance
4. Grounds maintenance
5. Security
6. Fire safety
7. Environmental safety
8. General safety
9. Telecommunications
10. Clinical engineering
11. Medical equipment management
12. Energy management
13. Technology evaluation and acquisition
14. Space management
15. Renovation and remodeling
16. Facilities planning
17. Disaster preparedness
18. Compliance management

---

[a] From McLarney VJ. Facilities Management In: Wolper LF, Pena JJ (eds). Health Care Administration: Principles and Practices. Rockville: Aspen Publishers, 1987: p 367.

operational reliability, to assess the special risks, and to respond to failure of utility systems that support the patient care environment.

The critical care nurse manager can integrate these organizational standards for plant, technology, and safety management at a unit level as well as many of the functional areas of facilities planning by putting into operation the structure standards of the American Association of Critical-Care Nurses (19). These standards are delineated in the Standards for Nursing Care of the Critically Ill (19). The comprehensive standards specific to facilities management include:

I. Comprehensive Standard: The critical care unit shall be designed to ensure a safe and supportive environment for critically ill patients and for the personnel who care for them.
II. Comprehensive Standard: The critical care unit shall be constructed, equipped, and operated in a manner that protects patients, visitors, and personnel from electrical hazards.
III. Comprehensive Standard: The critical care unit shall be constructed, equipped, and operated in a manner that protects patients, visitors, and personnel from fire hazard.
IV. Comprehensive Standard: The critical care nurse shall have essential equipment, services, and supplies immediately available at all times.

Although facilities management is primarily a support function for the critical care unit, a properly managed facilities management service working closely with the critical care manager can provide significant opportunities to minimize operating costs, and ensure necessary levels of safety and security for patients, families, visitors, and personnel (23).

BIOMEDICAL SERVICES

Biomedical engineers and technicians provide ongoing review of the safety, maintenance, and operation of electronic equipment (25). An inclusive biomedical management program encompasses the following functional program areas:

1. Preventive maintenance program;
2. Equipment evaluation and testing program;
3. Standards compliance program;
4. Staff training program for equipment operation;
5. Product recall management;
6. Equipment inventory and control;
7. Liaison with medical staff;
8. Incoming equipment and inspection (23).

Biomedical services range in scope from contracting for services, providing an external medical equipment maintenance organization, and sharing services to supporting all but the most technically advanced imaging equipment (23). Critical care units are particularly dependent on sophisticated diagnostic, therapeutic, and monitoring equipment. The critical care nurse manager can utilize the clinical engineering functions described by Shaffer et al. (26) as appropriate consulting guidelines for biomedical services (Table 7.15).

It is essential that the biomedical services personnel and critical care nursing staff work well together. The critical care nursing staff must be able to contact biomedical services for essential equipment 24 hours a day. The nursing staff members need to be able to trou-

Table 7.15. Consultation Guidelines for Biomedical Services[a]

1. Development and integration of new systems
   Planning
   a. System concept and design
   b. Facility, equipment, and interface diagrams
   c. Manufacturing and test specifications
   d. Cost estimates
   e. Operational and maintenance procedures
   Purchasing
   a. Sales literature files
   b. Sales quotations
   c. Buying decisions
   New installation
   a. Contractor liaison
   b. On-site installation and checkout support
   Training
   a. Scheduled nurse/technician training courses
   b. Educational seminars
   Evaluation
   a. Systems performance
   b. Statistics
   c. Cost-effectiveness
2. Operation, maintenance, and calibration
   Alignment and calibration
   Preoperation preparation checkout
   Routine performance and safety checks
   Equipment operation
   Failure repairs
   Incoming quality control inspection and test
   Spare parts inventories
   Schematic, instruction book, and reference library
   Operational improvement
3. Medical research and development support
   Proposal development
   New equipment design and construction
   Model shop operation
   Evaluation testing

[a] From Shaffer MJ, Carr JJ, Gordon M. Clinical engineering—an enigma in health care facilities. Hospital and Health Service Administration 1979;24(3):81.

bleshoot equipment malfunctions in the clinical setting regularly and with confidence. When problems arise with equipment that the nurse cannot resolve, biomedical services should be notified immediately for assistance. Biomedical services personnel depend upon the ability of the critical care staff nurse in the clinical setting to articulate the nature of the problem or equipment failure, because it is often difficult to duplicate the problem once the equipment has been removed from the clinical area. It is often helpful to maintain a biomedical log book in the critical care unit for routine maintenance or recording of nonemergent problems. Only with the ongoing collaborative support from biomedical services can the critical care nurse ensure a technologically safe and supportive environment for the critical care patient.

ENVIRONMENTAL SERVICES AND LAUNDRY

Maintenance of a clean, safe, and aesthetically pleasing environment is an important aspect of the provision of quality health care (27). Failure to provide such an environment may very well have such a negative effect on patients, visitors, and staff that they will choose not to return to a hospital that has such poor environmental services (27). The primary functional areas in environmental services are found in Table 7.16 (27).

The provision of a clean, safe environment for patients, staff, and visitors is the primary function of environmental services in the critical care unit (27). Critical care patients are more susceptible to pathogens because of the level of illness that causes a lowering of normal body defenses (28). Patients with a prolonged length of stay often require extensive invasive monitoring and experience a decreasing nutritional status, which contributes to a

Table 7.16. Functional Areas in Environmental Services[a]

1. Cleaning and disinfecting of all areas of the facility
2. Discharge unit cleaning
3. Carpet and upholstery maintenance
4. Window cleaning or monitoring of contract for services
5. Pest control or monitoring of contract for services
6. Monitor contract for services waste removal services, including infectious waste and chemotherapy waste
7. Room setups in conference rooms and auditoriums
8. Furniture relocation
9. Interior design
10. Laundry
11. Plant care

[a] From Kurth JM. Environmental services. In: Wolper LF, Pena JJ (eds). Health Care Administration: Principles and Practices. Rockville: Aspen Publishers, 1987: p 378.

general lowering of resistance to infection (28). Infection sources may be autogenous (the patient's own organisms), environmental, or a result of cross-infection (29,30). Regardless of the source, effective environmental services can reduce the risk of infection for critical care patients that would inevitably lead to prolonged hospitalization for the critical care patient.

Effective and efficient environmental services are easily identified. The hallmarks of a good environmental services department are delineated in Table 7.17 (27).

Laundry services may be provided inhouse or may be contracted out. The JCAHO (31) regulations stipulate that regardless of how laundry services are provided, an adequate supply of clean linen must be maintained, clean linen must be delivered in such a way as to minimize microbial contamination from surface contact or airborne deposition, and soiled linen must be collected in a manner that minimizes microbial dissemination into the environment.

Continuous and comprehensive environmental services are an essential support service for the critical care unit. Not only can inadequate environmental services adversely affect patient care outcomes, but in a competitive health care environment where the patients, staff, and visitors choose where they seek care and work, good environmental services can positively affect patient census, staff turnover, and organizational competitive viability (27).

INFORMATION SYSTEMS

There are three primary types of information systems in the hospital and critical care environment (32). These include clinical or medical information systems involved in the organized processing, storage, and retrieval of information to support patient care activities in the hospital; administrative information systems designed to assist in carrying out administrative support activities of the hospital such as finance; and fully integrated or total hospital information systems serving both clinical and administrative needs through centralized information storage and retrieval and with data communication links (33). Some characteristic functions of clinical information systems appear in Table 7.18 and functions of administrative information systems appear in Table 7.19.

The American Association of Critical-Care Nurses (AACN) has identified the importance of data acquisition and dissemination in AACN's Purpose, Long Range Goals, and Intermediate Strategies (34). Information systems can enhance the fulfillment of data acquisition objectives in critical care to:

1. Identify new knowledge and trends that influence critical care nursing in a timely manner;
2. Identify priorities for critical care nursing;
3. Examine outcomes of critical care nursing practice; and
4. Develop mechanisms to acquire and analyze data (34).

Table 7.17. Hallmarks of a Good Environmental Service[a]

1. The critical care unit is clean and pleasant-smelling
2. There are few valid complaints from patients, staff, and visitors about the environmental cleanliness
3. There are written policies and procedures for:
    —cleaning in surgery, isolation, and special areas such as the critical care unit;
    —care and use of equipment;
    —waste removal practices, i.e., infectious or hazardous waste;
    —safety practices, including safety data sheet pertaining to the products used in environmental services;
    —detailed general cleaning procedures describing responsibilities, frequency, and outcome
4. Employees are knowledgeable about environmental safety, policies and procedures, and feel a sense of commitment to the hospital and unit
5. There is an effective quality assurance program in place

[a] From Kurth JM. Environmental services. In: Wolper LF, and Pena JJ. (eds). Health Care Administration: Principles and Practices. Rockville: Aspen Publishers, 1987: p 381–385.

Table 7.18. Functions of Clinical Information Systems[a]

1. Computerized documentation of care including:
    - admission process
    - initial assessment, history, and physical
    - clinical laboratory study reports
    - radiological/diagnostic study reports
    - consultation reports
    - physician order entry for diagnosis and treatment
    - ongoing treatment and progress reports
    - nursing interventions and care
    - discharge diagnosis and summary reports
2. Computer-aided diagnosis
    - system collects patient data and communicates data from the patient record to user
    - system compares patient-specific information with available medical knowledge to aid in diagnostic decision-making
3. Computer-aided treatment and follow-up
    - computer generated treatment protocols and reminder/cue systems
    - radiation treatment planning systems
    - computerized patient follow-up systems for automated computer-generated patient notices
    - research data base generation
4. Patient monitoring systems
    - physiological parameter monitoring and trending
5. Laboratory automation and reporting
    - laboratory test requisition
    - scheduling of specimen collection and processing
    - result recording and reports
    - period summary reports
    - laboratory statistical reports
    - record-keeping for quality control and administrative control of laboratory operations
6. Medical record indexing and retrieval
7. Pharmacy information systems for
    - ordering
    - stock control
    - distribution/allocation
    - costing
    - drug information data base

[a] From Austin CJ. Information Systems for Hospital Administration. Ann Arbor: Health Administration Press, 1979: p 161–184.

Information systems can assist the critical care manager and nurse to:

1. Provide information on the impact of critical care nursing to critical care nurses, consumers of critical care nursing, and targeted other publics;
2. Share innovative and creative strategies that promote the value of professional nursing in an environment of increased competition in health care; and
3. Develop and promote utilization of new technologies for data acquisition, management, and dissemination (34).

Information systems are a useful tool that can enhance the critical care unit. The critical care patient, health care team, and manager must take an active role, however, in the generation of accurate data, utilization of information systems to enhance patient care outcomes, and establishment and exercise of individual information rights to ensure confidentiality if information systems are to be applied optimally in patient care (35).

LABORATORY AND RADIOLOGY

In providing patient care to critically ill patients, the critical care nurse is routinely involved with laboratory testing and diagnostic procedures (36). Nursing involvement may include multiple aspects of preparing for and assisting with the test or procedure, monitoring the patient for effects during or after the test or procedure, interpreting and reporting

Table 7.19. Functions of Administration Information System

1. Facility utilization and scheduling systems
    —utilization review
    —monitor inpatient activity
    —monitor clinic and emergency room activity
    —utilization of individual services
    —advance patient booking and scheduling
    —report generation
2. Financial information systems
    —payroll preparation and accounting
    —accounts payable processing
    —collecting charges, patient and third party billing, accounts receivable processing
    —general ledger accounting
    —cost allocation system for proration of costs for nonrevenue-generating activities and overhead
    —financial report preparation
    —budgeting and cost control
3. Materials and facilities management systems
    —computerized purchasing services
    —inventory control systems
    —computerized menu planning and food service management
    —monitoring of preventative maintenance
    —energy management
    —project scheduling and control
4. Personnel data systems
    —maintaining and updating employee records
    —retrieving file information
    —position control
    —labor analysis reports
    —personnel problem analysis
    —inventory of employee special skills and certifications
    —labor cost allocation reports
    —employee productivity and quality control reports
    —scheduling

[a] From Austin CJ. Information Systems for Hospital Administration. Ann Arbor: Health Administration Press, 1979: p 185–204.

the test or procedure results, and educating the patient and family about the test or procedure (36).

Pathology and medical laboratory services and consultation required by the JCAHO (37) include anatomic pathology, hematology, chemistry, microbiology, clinical microscopy, parasitology, immunohematology, serology, virology, and nuclear medicine. The radiology department provides acquisition, manipulation, storage, retrieval, and analysis of images that have been increasingly shifting from analog to digital formats over the last decade (38). Common radiological modalities used in the diagnosis and treatment of critical care patients include x-ray, computerized tomography, digital subtraction angiography, ultrasound, positron emission tomography, magnetic resonance imaging, and nuclear medicine (38). Responsibilities of the critical care nurse in laboratory testing or procedures are found in Table 7.20 (36,39).

Often the laboratory and radiology services provide a policy and procedure book for nursing units to assist staff members in facilitating the test or procedure. Despite the availability of such policies and procedures, there are often problems that arise among the laboratory and radiology services and the critical care unit staff members related to obtaining appropriately sampled and labeled specimens, timing of tests, patient transport and monitoring, and result reporting. It is often helpful for the critical care manager to meet regularly with key laboratory and radiology service managers to discuss and resolve such issues.

Table 7.20. Responsibilities of the Critical Care Nurse in Laboratory Testing and Radiology or Diagnostic Procedures[a]

1. Understands the purpose of the test
    —specific indications for the test
    —relevant physiology pertaining to understanding of the test
    —the clinical importance of the test
2. Patient preparation
    —necessary physical and psychological preparation
    —diet or drug restriction
    —test or procedure description
    —ensure that the necessary informed consent or permissions are obtained before test or procedure
3. Ensures that the necessary equipment is available
4. Competent in nursing responsibilities required during the test or procedure
5. Ensures necessary precautions are taken during the test or procedure and is aware of factors that interfere or invalidate the test or procedure results
6. Interprets normal and abnormal results
7. Provides post-test or postprocedure monitoring and care

[a] From Leavelle DE. Forward. In: Diagnostics: Patient Preparation, Interpretation, Sources of Error, Post-Test Care. Springhouse: Intermed Communications, 1983:20–21.

COMMUNICATION

Effective communication is important for critical care managers because communication is the necessary process by which the management functions of planning, organizing, directing, and controlling are accomplished, and communication is the single activity to which managers devote an overwhelming account of time (40). Critical care managers are involved in interpersonal communication that occurs when information and meaning are exchanged from one person to another or in small groups of people (40). They are also involved in organizational communication or the process by which the critical care nurse manager uses the established system to receive and relay information to people or departments within the organization as well as to relevant individuals and groups outside it (40).

The advantages to formal communicaton include predictability, documentation, control over the message, and that it is verifiable (41). Advantages to informal communication include flexibility, speed, availability of immediate feedback, and that it is more personal (41).

Organizational communication contributes to critical care patient outcomes because critical care delivery is dependent on multidisciplinary health care providers, ancillary services, support services, and diagnostic services. It is appropriate to involve these disciplines in patient care goal setting, achievement, and evaluation through an innovative model of care delivery such as managed care.

MANAGED CARE

Managed care begins with a description of the processes and outcomes involved in the care of specific groups of patients, such as patients experiencing a myocardial infarction or coronary artery bypass graft (3). Physicians, nurses, and other health care providers identify the processes and outcomes in terms of a standard critical pathway or managed care plan (3,42). The group determines the key incidents that must occur for the specific group while considering reasonable length of stay timeframes and efficient resource utilization (42). The managed care plan articulates patient outcomes in light of the therapeutic process and available resources. The managed care plan is a template for a pattern of preferred care defined by the group of providers if the system functions well and if the patient experiences no complications. The appropriate managed care plan is selected when the patient is admitted, individualized, and modified as needed for the particular patient's case by the physician, nurse, and other providers (3,42). The managed care plan or pathway is reviewed regularly during the provision of care by all health care providers to identify progress and patient variances (3,42). Variances are expected and are addressed by the provider, ancillary, or support service accountable for the patient outcomes affected by the variance (3,42).

Bower (3) described many benefits of managed care including:

1. Patients participate more in their care because they have a clear direction, are more aware of their progress, and have more insight into their care.
2. Practice becomes more patient outcome-oriented.
3. Length of stay is controlled and often reduced.
4. Resource utilization is controlled and often reduced.
5. Staff nurses experience a greater sense of control and satisfaction regarding the care of patients.
6. Orientation of new staff is facilitated because standards of care are delineated in a usable manner.
7. Collaboration among health care providers is increased.
8. Standards of care are developed for specific case types.
9. Contributions of intermittent staff are more outcome-oriented.

Client-centered conferences enhance the managed care delivery process in terms of addressing problems and variances. Some relevant considerations the critical care manager should remember in facilitating client-centered conferences include:

1. Adequate preparation by the critical care nurse manager or case manager for the conference will achieve the conference expectations.
2. Involvement of appropriate providers and services will enhance multidisciplinary care effectiveness.
3. Prior announcement of time, place, purpose, and duration of the conference promotes group preparation, attendance, and focusing of the group on the conference purpose.
4. Interaction of conference members on an equal basis encourages active participation and leads to usable solutions to the problems.
5. Identifying specific problems to be considered expedites the formulation of interventions.
6. Analyzing the critical care delivered in light of established goals enables members to validate their behavior and to devise ways to improve care (40).

The focus of managed care is on redesigning tools and systems to promote outcome-oriented, fiscally responsible care and is an exciting challenge for critical care nurse managers (3). However, it will be the willingness and desire on the part of the entire health care team—the physicians, the ancillary services, and the support services—to break through traditional role barriers and to reach out to assist and support the critical care nurse in the delivery of managed patient care that will determine the success of this innovative model (43).

REFERENCES

1. Clemmer TP, Orme JF. An integrated approach to the patient with acute respiratory failure. In: Clemmer TP, Orme JF (eds). Critical Medicine. Salt Lake City: LDS Hospital, 1981: p 1–23.
2. Styles MM. Reflections on collaboration and unification. Image: The Journal of Nursing Scholarship 1984; 26(1):21–23.
3. Bower KA. Managed care: controlling costs, guaranteeing outcomes. Definition: The Center for Nursing Care Management 1988;3(3):1–3.
4. McClure ML, Nelson MJ. Trends in hospital nursing. In: Aiken LH (ed). Nursing in the 1980s: Crises, Opportunities, Challenges. Philadelphia: JB Lippincott, 1982: p 59–74.
5. ASHP. ASHP statement on the pharmacist's clinical role in organized health-care settings. ASHP Position Statement, November, 1988.
6. Kubica AJ, Poremba AC. Pharmacy. In: Wolper LF, Pena JJ (eds). Health Care Administration: Principles and Practice. Rockville: Aspen Publishers, 1987: p 456–470.
7. ASHP. Practice Standards of the American Society of Hospital Pharmacists. Bethesda, MD: American Society of Hospital Pharmacists, 1984:11.
8. JCAHO. The Joint Commission 1990 Accreditation Manual for Hospitals. Chicago: Joint Commission on Accreditation of Healthcare Organizations, 1989: p 225–233.
9. Thompson DF. Attitudes of pharmacists and nurses toward their responsibilities in the drug treatment process. AJHP 46(Feb):257.
10. Kenner CV. Respiratory assessment and basic ventilatory care. In: Kenner CV, Guzzetta CE, Dossey BM (eds). Boston: Little, Brown and Company, 1985: p 217–258.
11. Comroe JH. Physiology of Respiration, 2nd ed.

Chicago: Year Book Medical Publishers, 1974: p 275.
12. Smoker JM. Personal correspondence. York, PA: November 14, 1989:1–3.
13. JCAHO. The Joint Commission 1990 Accreditation Manual for Hospitals. Chicago: Joint Commission on Accreditation of Healthcare Organizations, 1989: p 177–192.
14. JCAHO. The Joint Commission 1990 Accreditation Manual for Hospitals. Chicago: Joint Commission on Accreditation of Healthcare Organizations, 1989: p 181–182.
15. Campbell SM. Nutrition. In: Kinney MR, Packa DR, Dunbar SB (eds). AACN's Clinical Reference for Critical Care Nursing, 2nd ed. New York: McGraw-Hill Company, 1988: p 334–371.
16. Butterworth CE. The skeleton in the hospital closet. Nutrition Today 1974;9:4–8.
17. JCAHO. The Joint Commission 1990 Accreditation Manual for Hospitals. Chicago: Joint Commission on Accreditation of Healthcare Organizations, 1989: p 17–26.
18. JCAHO. The Joint Commission 1990 Accreditation Manual for Hospitals. Chicago: Joint Commission on Accreditation of Healthcare Organizations, 1989: p 186.
19. Sanford SJ, Disch JM. AACN Standards for Nursing Care of the Criticall Ill, 2nd ed. Norwalk: Appleton and Lange, 1989: p 15–46.
20. Crittenden FJ. Working with staff alliance. Discharge planning for health care facilities. Los Angeles: University of California Extension Allied Health Publications, 1983: p 39–70.
21. AHA. Discharge planning update: an interdisciplinary perspective for health professionals. Chicago: AHA, 1980;1(1):24.
22. Crittenden FJ. Philosophy, goals, and tools for working with patient and family. Discharge planning for health care facilities. Los Angeles: University of California Extension Allied Health Publications, 1983: p 71–86.
23. McLarney VJ. Facilities management. In: Wolper LF, Pena JJ (eds). Health Care Administration: Principles and Practices. Rockville: Aspen Publishers, 1987: p 366–378.
24. JCAHO. The Joint Commission 1990 Accreditation Manual for Hospitals. Chicago: Joint Commission on Accreditation of Healthcare Organizations, 1989: p 195–202.
25. Gaylin W. The ethical, legal, and technological setting. In: Rakich JS, Longest BB, Darr K (eds). Managing Health Service Organizations, 2nd ed. Philadelphia: WB Saunders, 1985: p 80–141.
26. Shaffer MJ, Carr JJ, Gordon M. Clinical engineering an enigma in health care facilities. Hospital and Health Service Administration 1979;24(3):81.
27. Kurth JM. Environment services. In: Wolper LF, Pena JJ (eds). Health Care Administration. Rockville: Aspen Publishers, 1987: p 378–385.
28. Trought EA. Safety. In: Kinney MR, Packa DR, Dunbar SB (eds). AACN's Clinical Reference for Critical-Care Nursing, 2nd ed. New York: McGraw-Hill, 1988: p 83–95.
29. Kottra CJ. Infection in the compromised host-an overview. Heart Lung 1983;12(10):10–14.
30. Levenson SM, Laufman H. Infection hazards of surgical intensive care. In: Levenson L, Thomson CW (eds). Manual of Surgical Intensive Care. Philadelphia: WB Saunders, 1977: p 157–160.
31. JCAHO. The Joint Commission 1990 Accreditation Manual for Hospitals. Chicago: Joint Commission on Accreditation of Healthcare Organizations, 1989: p 69.
32. Austin CJ. Information systems to support direct patient care: clinical information systems. Information Systems for Hospital Administration. Ann Arbor: Health Administration Press, 1979: p 161–184.
33. Austin CJ. Information systems to support administrative operations in hospitals. Information Systems for Hospital Administration. Ann Arbor: Health Administration Press, 1979: p 185–204.
34. AACN. AACN's purpose, long range goals, and intermediate strategies. AACN Position Statement, February, 1988.
35. Patrikas EO, Liebler JG. Clinical information services. In: Wolper LF, Pena JJ (eds). Health care administration. Rockville: Aspen Publishers, 1987: p 399–411.
36. Leavelle DE. Foreword. In: Diagnostics: Patient Preparation, Interpretation, Sources of Error, Post-Test Care. Hickville: Springhouse Intermed Communications, 1983: p 20–21.
37. JCAHO. The Joint Commission 1990 Accreditation Manual for Hospitals. Chicago: Joint Commission on Accreditation of Healthcare Organizations, 1989: p 137–160.
38. Hamilton RJ, James AE, Stephens WH, Pendergrass HP. Radiology. In: Wolper LF, Pena JJ (eds). Health Care Administration. Rockville: Aspen Publishers, 1987: p 445–453.
39. Storlie FJ. Overview: new diagnostic challenges. In: Diagnostics: Patient Preparation, Interpretation, Sources of Error, Post-Test Care. Springhouse: Intermed Communications, 1983: p 22–25.
40. Douglass LM, Bevis EO. Predictive principles of effective communication: interpersonal and group. Nursing Management and Leadership in Action, 4th ed. St. Louis: CV Mosby, 1983: p 169–233.
41. Levson E, Guy M. Communicating. In: Cardin S, Ward CR (eds). Personnel Management in Critical Care Nursing. Baltimore: Williams & Wilkins, 1989: p 78–91.
42. Woldrum K. Critical paths. Definition: The Center for Nursing Care Management. 1987;2(3):1–4.
43. Tonges MC. Redesigning hospital nursing practice: the professional advanced care team (ProACT) model, part 1. J Nurs Adm 1989;19(7):31–38.

Chapter 8

Integrating Materiel Support Systems

LAWRENCE A. DAVIDSON,
JOAN GYGAX SPICER

The goal of the nurse manager in relation to materiel support for the critical care unit is to integrate the processes of materiel support with the daily operations of the critical care unit in order to achieve the highest quality service for the nurse at the bedside delivering nursing services. This can be accomplished through a synergistic approach to the design, implementation, development, and control of materiel delivery systems.

PLANNING FOR QUALITY SYSTEMS

"Organizational systems and subsystems must be goal oriented, and the process of systems planning requires the development of synergistic goals towards which planning methodology is directed" (1). Therein lies the imperative for developing clinical area-support service goals. The primary goal for the critical care unit and the materiel management department should be to provide the highest quality of care, goods, and services at the best total cost. Specifically, the goal is to provide the critical care unit with a total supply system in which the right supply is always available at the right place and time, in the right quantities, in the right condition, and at the right cost.

The nurse manager needs and, in fact, should want a system design that is both observable and measurable in terms of quantifiable patient care outcomes. The system should be standards-driven and prevention-oriented. The standard for patient care and support services to patient care units should be ZERO-DEFECTS, which means that every employee, whether clinical or ancillary, *DOES IT RIGHT THE FIRST TIME.*

SYSTEMS ANALYSIS, DESIGN, AND IMPLEMENTATION

When the word "system" is used in business it defines a group of elements or parts that are integrated around the common purpose of achieving some identifiable goal (2). Each element of a system must be integrated in such a manner that it works toward achieving systems goals, rather than individualized specific goals, and, although not every part of a system must necessarily work together, every part must have some type of logical relationship. Systems are comprised of subsystems; in the case of materiel support to the critical care unit, these systems and subsystems are integrated with the greater hospital system.

The design and development of a quality, self-perpetuating system requires the involvement of all users of the system. According to Sanderson (2), a system's life cycle includes four distinct phases: the conceptual phase, the analysis and design phase, the implementation phase, and the operational and control phase. This life cycle is depicted graphically in Figure 8.1. Sanderson delineates the criteria required for each phase.

Conceptual Phase

1. Identification of existing needs or potential deficiencies within existing systems;
2. Establishment of system concepts that provide initial planning to overcome potential or existing deficiencies;
3. Identification of initial economic, technical, and environmental factors that will enhance or detract from the system's practicality;
4. Identification of those areas of the system

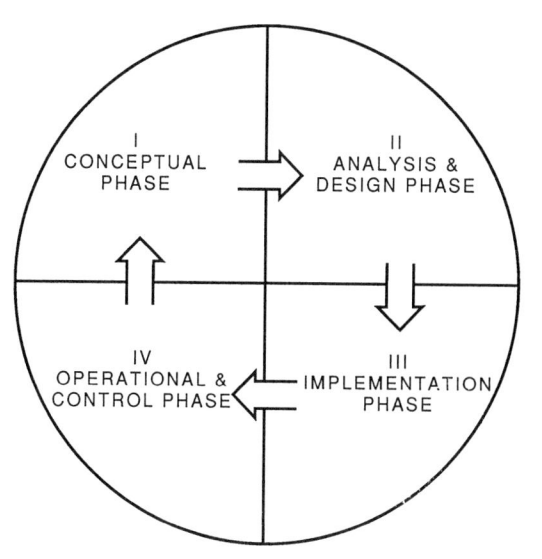

Figure 8.1. A system's life cycle. (From Sanderson ED. Effective Hospital Materiel Management. Rockville: Aspen Systems Corporation, © 1985.)

where high risk or uncertainty exists, and delineation of plans for further exploration of these areas;
5. Identification of necessary support subsystems;
6. Preparation of the documentation required to support the system such as policies, procedures, job descriptions, and a budget;
7. Development of implementation schedule.

Analysis and Design Phase
1. Final identification of the required economic resources;
2. Identification of the final system performance requirements;
3. Preparation of detailed plans required to support the system;
4. Identification of those areas of the system where high risk or uncertainty exists, and delineation of plans for further exploration of these areas;
5. Identification of necessary support subsystems;
6. Preparation of the documentation required to support the system, such as policies, procedures, job descriptions, and a budget;
7. Development of implementation schedule.

Implementation Phase
1. Final updating of plans conceived and defined in preceding phases;
2. Verification of systems requirements and objectives;
3. Final preparation and transmission of all policy and procedures;
4. Development of final plans to support system during operational phase;
5. Beginning of implementation.

Operational and Control Phase
1. Utilization of the system by intended user;
2. Actual integration of the system into existing organizational systems;
3. Evaluation of the operational sufficiency of the system;
4. Evaluation of the adequacy of supporting subsystems or other systems;
5. Provision of feedback to administration and others concerned with system planning and development (2).

Establishing Goals and Objectives

It is necessary for all users and providers in the system to identify and establish mutually beneficial goals and objectives if the integrity of the system is to be maintained. Mutual and active participation by both users and providers serves to establish a kind of win-win philosophy, in which both users and providers are directed toward the realization of the primary operational goal, and both users and providers experience the positive outcomes of patient care.

It is important to take a synergistic approach to the design and development of materiel support systems to the clinical area. If materiel services personnel, with good intentions, design a system to support the patient care activities without the involvement of nursing, the nurses do not have the vested interest in seeing the system succeed. When the planning includes both provider and user, the collaborative nature of the effort generates the tolerance and patience that underlie good working relationships. When designing materiel support systems to the critical care unit, as a minimum, these individuals should be included:

1. The nurse manager of the critical care

unit(s) and the director of materiel services; these individuals have ultimate responsibility and accountability for the system.
2. The medical director(s) of the unit(s); participation by the physicians reaffirms the organization's commitment to the system. If the medical staff participates, the materiel services area is able to establish a personal interface with the physicians, thereby increasing access to users of the system who, both historically and professionally, are somewhat detached from materiel services personnel.
3. Relevant nursing and general services administrators; participation by these individuals, if even merely in a supportive role, reaffirms the organization's commitment to the system.
4. Relevant commodity managers; those individuals having responsibility and accountability for *facts* of the system should participate in the design of the system. These individuals might include:
   • Director of Pharmacy Services;
   • Director of Linen Services;
   • Director of Respiratory Therapy Services;
   • Distribution and Purchasing Managers;
   • Central Sterile Processing Manager.
5. Shift supervisors; these individuals, both clinical and ancillary, are responsible for maintaining the integrity of the system on a 24-hour basis.
6. Nonsupervisory nursing and materiel services personnel; involvement of these individuals will encourage and establish a personal interface and relationship among individuls from the various departments in the system. Nonsupervisory personnel, who participate in the system design process, develop a sense of accountability and responsibility for the system and often convey an air of optimism and excitement to their peers.

Establishing Standards and Specifications

In the 1983 *Summary Report and Recommendations,* the National Commission on Nursing made the following recommendation, "The presence and quality of supporting services to the patient care unit is a major determinant in the effectiveness of the delivery system and the satisfaction of the professionals working in the system. To assure a high quality of support services, nursing, as part of the management team, should actively participate in establishing standards for, and evaluating the quality of support services for patient care" (3). The American Association of Critical Care Nurses (AACN) has established standards in relation to the provision of materiel to the critically ill patient (4). The specific standards appear in Appendix A.

Webster defines standards as "something set up as a rule for measuring, or a model to be followed," and define specification as "a description of work to be done, and materials to be used." Standards and specifications are the terms and conditions by which operational endeavors are to be executed. Standards and specifications are prenegotiated parameters and are directed toward the achievement of well-defined goals. Standards and specifications must be observable and quantifiable against patient care outcomes, if one is to offer objective statements concerning goals and objective attainment. That is, as an indication of his or her success and control, the manager must be able to document and quantify departmental endeavors.

During the developmental process it is important to avoid vagueness or ambiguity and to establish clearly defined standards and specifications. The following examples help elucidate this fact.

1. *Poorly defined standard/specification;* the distribution technician will make continuous rounds throughout the critical care units to stock supplies, remove soiled instruments, and retrieve documents.
   Appropriate standard/specification; the distribution technician will conduct rounds of the critical care units to include the surgical intensive care unit, coronary care unit, neonatal intensive care unit, and the emergency room during the following intervals:
   • 3:15–3:30, surgical intensive care unit;
   • 3:35–3:50, coronary care unit;
   • 4:00–4:15, neonatal intensive care unit;
   • 4:25–4:40, emergency room.
   The technician will deliver requested sup-

plies to the exchange cart, place requested documents in the receiving basket at the nurse's station, and remove soiled instruments from the soiled utility room. Before leaving the unit, the technician will consult the unit secretary for further requests.

2. *Poorly defined standard/specification;* the unit secretary will contact the central service department as soon as possible after a Code Blue event and request a replacement crash cart.

   Appropriate standard/specification; upon completion of the Code Blue effort, the staff nurse will direct the unit secretary to request a replacement crash cart. The unit secretary will contact the central service department within 5 minutes and request a replacement crash cart. A central service technician will deliver the crash cart to the unit within a 10-minute timeframe. Before leaving the unit, the central service technician will inform the unit secretary that the crash cart has been exchanged, and that he or she is in possession of all applicable charge documents.

In contrast to the poorly defined standards/specifications, the appropriate standards/specifications accomplish the following objectives:

1. Delineate and clarify responsibility and accountability;
2. Establish procedures for accomplishing tasks;
3. Establish quantifiable performance criteria;
4. Provide an opportunity to establish target-area control tools for monitoring the system;
5. Provide an opportunity for establishing an objective format for reporting the success and/or failure of the system.

Delineated below are minimum areas of standards/specifications to be developed for the materiel support system. These include standards/specifications for:

- Delivering supply carts, including:
  —Medical/surgical exchange carts,
  —Linen carts,
  —Isolation carts,
  —Dietary carts,
  —Case carts, crash carts, and other specialty carts;
- Replenishing par-level storage areas;
- Providing durable patient care equipment;
- Providing sterile instruments and sterile instrument packs;
- Replacing medical gases;
- Procuring, delivering, and maintaining specialty beds and support equipment;
- Responding to routine and stat requests;
- Confirming receipt of stock and nonstock requests;
- Delivering and expensing stock and nonstock requests;
- Conducting continual service rounds;
- Analyzing and reporting weekly inventory usage rates and costs;
- Adjusting inventory levels to support usage data;
- Performing patient charge capture procedures;
- Performing lost patient charge procedures;
- Aerating of ethylene oxide-sterilized materiel;
- Performing emergency recalls of supplies, equipment, and instrumentation;
- Stocking and rotating sterile supplies;
- Containing and transporting soiled instruments and equipment;
- Containing and transporting infectious waste;
- Responding to internal and external disasters;
- Adhering to dress code standards;
- Adhering to professional protocol within the units;
- Completing and submitting stock and nonstock requests;
- Completing and submitting work requests;
- Adhering to protocols for telephone requests;
- Adding or deleting inventory items;
- Performing product substitutions;
- Performing order entry and manual requisition procedures;
- Providing and updating central storeroom and central service department catalogs;
- Obtaining supplies, instrumentation, and equipment after normal central stores/central service department hours of operation;
- Lending and/or borrowing supplies, instru-

mentation, and equipment to or from other institutions;
- Care of supplies, instrumentation, and equipment on the nursing unit;
- Cleaning and disinfecting supply carts and equipment.

Standards and specifications, like policies and procedures, change with system revisions and should be revised accordingly. Operational standards and specifications should be treated as an integral component of the new employee orientation protocol, and discussion of them should be part of the continual inservice educational program. It is vital that clear delineation and dissemination of responsibility, accountability, and authority occur.

Policies and Procedures

The efficient and effective materiel services department is guided by 100–150 policies and procedures, a significant amount of which are directly related to materiel services to the patient care units. Those materiel services policies and procedures that directly affect the critical care units should be addressed in the critical care nursing policies and procedures as well. Because of their interactive nature, these policies and procedures must be developed collectively by both the user and provider of services. To be effective tools, policies and procedures must be readily accessible to all nursing personnel. Policies and procedures are institutionally dependent, but, as an absolute minimum, the areas delineated in the standards and specifications section should be included within the materiel services policies and procedures for the critical care unit.

Internal Communication

Managers are required to become increasingly aggressive in controlling the total costs of supplies, equipment, and services, while still providing the highest quality of patient care. Effective communication systems contribute significantly to the financial bottom line. Among the results of poor communication or miscommunication specific to materiel systems are:

- Quality problems with goods and services;
- Obsolescence;
- Expensive rework;
- Inventory replication;
- Lost opportunity (so-called opportunity costs);
- Inter- and intradepartmental factionalism;
- Detrimental or negative patient care outcomes.

Successful organizations are redefining and refining communication systems. Shifting from hierarchical communication structures to matrix communication structures consisting of lateral relationships within the organization. Matrix communication systems, see Figure 8.2, focus on the leadership potential of all

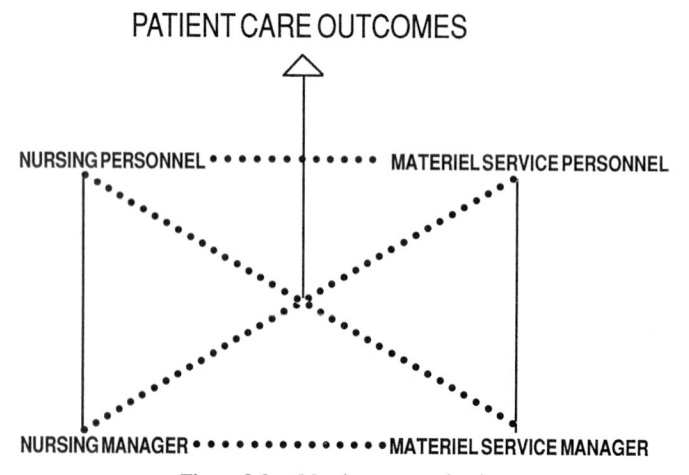

Figure 8.2. Matrix communication.

employees and accomplish, among other things, the following objectives:

- The development of mutual goals;
- The elimination of historical inter- and intradepartmental power struggles;
- The development of multilateral educational experiences.

When designing or refining the materiel support system for the critical care unit, it is important that a communication matrix is developed and directed toward the achievement of interactive goals. Both users and providers must participate if the integrity of the system is to be maintained. An ongoing, committee-type approach involving clinical practitioners and materiel services personnel is a beneficial method to improve support systems to the critical care units. The committee usually consists of the managers of each area, the various shift supervisors, and representatives from the nonsupervisory staffs. Users and providers working together can:

- Establish clearly defined communication flow patterns;
- Establish responsibility and accountability of the nursing department and materiel services department for maintaining the integrity of the systems;
- Establish mutual goals and objectives and expectations of the system stated in terms of quantifiable patient care outcomes;
- Provide a review of the ongoing performance monitoring of the system.

The approach allows the departments to establish a perpetual interface between users and providers, thereby enhancing interpersonal relationships and avoiding "exception only" or "crisis-oriented" communication. Much too often, materiel services personnel play a somewhat innocuous role in patient care delivery systems but being responsive to multitudinous demands, being merely marginally aware of the system, receiving admonishments for "inadequacy" during tense moments of poor performance, and being relegated to "invisibility" during the positive moments.

The committee-type approach is prevention-oriented allowing users and providers to discuss the successes and failures of the system in an objective form and to develop action plans for revising the system. According to Seiler et al. (5), communication is both interactional and transactional. Interaction is the exchanging of messages that occurs among the people involved in the communication process. Transaction carries the concept one step further in that it views communication between people as a simultaneous sharing event (5). An objective communication forum can serve to enhance interactive and transactive perspectives, thereby contributing to positive patient care outcomes.

Shared Educational Experiences

The design, implementation, or refinement of support systems to the critical care units necessarily requires an ongoing educational program. The nursing department managers and staff nurses should play a key role in the education of the materiel services staff. Conversely, the materiel manager and the materiel services staff can and should play key roles in the education of the nursing staff.

A shared experience in other departments should occur, at a minimum, during new employee orientation periods and should address the following areas:

- Departmental structure;
- Intradepartmental communication protocol;
- Goods and services offered;
- Financial criteria;
- Systems expectations and limitations;
- Problem-solving protocol;
- Daily operational requirements.

Shared experiences, like matrix communication systems, provide an opportunity to establish and refine personal relationships and foster a team-oriented perspective. Nurses can orient materiel services personnel to the clinical environment, an environment that is often unfamiliar and intimidating to materiel services personnel. Furthermore, nurses can assist with the development of relationships between physicians and materiel services personnel—relationships that are often nonexistent. Materiel services personnel can provide the following educational experiences to nurses:

- Product and equipment utilization techniques;

- Product and equipment maintenance techniques;
- Inventory control methodologies;
- Product standardization information;
- Product availability information;
- Space and materiel requirements planning (MRP) techniques.

Finally, in support of the matrix communication system mentioned previously, nursing and materiel services personnel should "take each other to the point of the problem." This means that when breaks in the integrity of the system occur, the individuals responsible should be able to avail themselves of immediate educational opportunities, a proactive, "hands-on" approach to educational endeavors. Problems that occur between nursing units and materiel support departments are often relegated to reactionary responses that rectify only the immediate situations and the possible causes are overlooked until a multitude of these problems finally overwhelms the system. Obviously, systems problems should be resolved in a systematic, objective manner. However, both nurses and materiel services personnel should avail themselves of the immediate educational opportunities of "point-to-point" problem resolution.

QUALITY CONTROL

For the industrial/manufacturing environment Philip B. Crosby (6) has defined the appropriate quality attitude as one that embraces the following four absolutes of quality:

1. The First Absolute: the definition of quality is conformance to requirements.
2. The Second Absolute: the system of quality is prevention . . . not appraisal.
3. The Third Absolute: the performance standard is zero defects.
4. The Fourth Absolute: the measurement of quality is the price of nonconformance (6).

Implementing a quality control program requires the manager to implement the standards/specifications, policy/procedures, communication, and education structures.

Establish the Performance Criteria of the System

The quality control program requirements should not burden an already busy staff. The program should be easy to use and should demonstrate conformance to requirements. The collection, analysis, and presentation of quality control data serves to describe patient care efforts as occurring in accordance with quantifiable standards and guidelines. The program should be more or less self-documenting in the sense that the activities of staff personnel should be recorded by the individuals themselves at the time of occurrence. To the nurse, this is a very obvious verity, and this endeavor occurs continuously in the form of nursing documentation. It also occurs continuously in the materiel services area in the form of such documentations as inventory/replenishment documents.

Determine Responsibilities for Maintaining System Integrity

Responsibility and accountability for maintaining system integrity must be unambiguously defined in both the nursing department and the materiel services department and for any intermediary users of the system on a 24-hour basis. Employees on each shift having this responsibility and accountability should be identified for both the nurse manager and materiel manager. This type of shared information serves to establish communication channels and to establish an efficient method for managing breaks in the system integrity.

Establish the Performance Monitoring System

The quality of a system is usually assessed and monitored by employing one, or all, of the following techniques:

1. Disseminating and analyzing departmental quality survey (customer service surveys);
2. Implementing quality assurance monitors in accordance with standard sampling plans;
3. Implementing statistical process controls with target area control documentation;
4. Implementing systematic, closed-loop type techniques for responding to incidents.

Quality Surveys

Quality surveys are questionnaires to which users of a system respond. The questions encompass facets of a department's operations, and the users of a service area's operations assess the strengths and weaknesses of the materiel services department's performance. The surveys can provide insight regarding a specific department's operations and can help to sustain a communication linkage between both users and providers in the system. The disadvantage of quality surveys is that they are opinion-oriented and therefore do not readily lend themselves to statistical validity. Also, more often than not, questionnaires are completed and submitted by only a small percentage of the employees/customers.

Quality Assurance Monitors in Accordance with Sampling Plans

Monitors are also a type of quality survey by which facets of the materiel services department's operations are assessed. Monitoring is conducted in accordance with an approved sampling plan, and are directed toward assessing a department's conformance to standards/specifications.

Quality assurance monitors, like quality surveys, allow the manager to assess a department's strengths and weaknesses, and to concentrate resources on areas of nonconformance. The shortcoming of monitoring methodologies are that they are reactionary in nature and, in this sense, are problem-oriented rather than prevention-oriented.

Statistical Process Control

Statistical process control (SPC) is also a type of auditing methodology, but one that uses target area control methodology as a prevention-oriented management tool. In the production environment, SPC is used to monitor production processes to ensure that products are manufactured within specified ranges before they are released to customers. Typically, a high, middle, or low value is specified, and the production process is monitored accordingly. Should the values remain within specified tolerances, the production process continues, and the products are released to consumers. Management is notified, however, when the middle range is exceeded, thereby allowing them to evaluate the process and execute corrective action before nonconformances occur.

With some inventiveness, the nurse manager can utilize target area controls to monitor support service department performances. This is especially true for the following areas:

- Medical/surgical supply storage carts and cabinetry;
- Durable patient care equipment provision;
- Sterilized instrument provision;
- Materiel services response times.

The distinct advantage of SPC is that it is prevention-oriented rather than problem-oriented. Another advantage is that nurse managers can avail themselves of an objective, quantifiable tool for assessing materiel support services. So, as long as the predetermined ranges are not exceeded, nurse managers can have confidence that support department services are being provided in accordance with predetermined standards and specifications.

Incident/Problem Resolution Techniques

Incidents that lead to the breakdown of a system's integrity must be responded to in a controlled, systematic manner. The response should be directed toward a systems approach to problem resolution and the prevention of similar future occurrences. As a minimum, the methodology should include the following parameters:

- Timeframe for the response(s) in accordance with prenegotiated standards and specifications;
- Delineation of the individuals responsible for responding;
- Method by which the response will be made.

A format for the response should include:

- A concise and clear description of the incident(s):
  —What happened?
  —When did it occur?
  —Where did it occur?
  —What caused the incident?
- A statement of the responding individual's understanding of the implication and/or effect of the incident(s);
- A concise outline of the resulting approach to prevent similar, future incidents;
- Delineation of policy and procedure revision, if applicable, with signature approval of all relevant department managers.

Usually a quality control-type document is utilized as a management tool for resolving incidents and/or problems. Although institutionally dependent, a representative document is presented in Figure 8.3.

MATERIEL SERVICES FEEDBACK FORM

FROM: _____ UNIT _____   Date: _____
         NURSE MANAGER
                                         Time: _____
TO: Director, Materiel Services
Please check appropriate information:

☐ **DISTRIBUTION**

☐ **EXCHANGE CARTS**   ☐ **DISPATCH**   ☐ **STOCKING - Area 1**
  ☐ items(s) missing   ☐ received incorrect item   Clean Utility
  ☐ inadequate amount of item(s) ☐ timeliness   ☐ not stocked
  ☐ wrong item(s) on cart   ☐ needed to send > 1   ☐ inadequate stocking
  ☐ time of arrival on unit   requisition to get item   ☐ incorrect stocking
  ☐ items in wrong location on cart ☐ other-describe below   ☐ other-describe below
  ☐ cart not exchanged since   Area 2
  ☐ other-describe below   ☐ not stocked
                                                                   ☐ inadequate stocking
☐ **PROCESSING**   ☐ incorrect stocking
  ☐ equipment   **OTHER**   ☐ other-describe below
  ☐ trays   ☐ _____
  ☐ dirty pickup

DETAILS OF INCIDENT OR OTHER COMMENTS (Including location and date):

_____
_____
_____
_____
_____
_____
_____

ACTION TAKEN BY MATERIEL SERVICES:

_____
_____
_____
_____

                                        Signature: _____

Figure 8.3. Materiel services feedback form. (Modified from: Materiel Services and Nursing Task Force, Materiels Services Feedback Form, The Medical Center at the University of California, San Francisco, 1989.)

It is important that both the nurse manager and the materiel services manager define and clarify early in the working relationship between the departments what constitutes an incident. An incident may be defined as when the results of some activity or omissions expose the patient to potential morbidity and/or mortality. This definition takes a broad view of an incident because there may not be a measurable effect of the activity or omissions. For this reason, it is rather obvious that both the definition and clarification of what constitutes a patient incident must occur early in the working relationship of the departments.

It is equally important that incidents originating in nursing performance are clearly defined early in the relationship of the departments. Since the materiel manager is not dealing directly with patients, the definition of an incident occurring in the material services department might include loss of man hours within the department because nursing personnel did not follow through on their commitment to maintain the integrity of a system as established during the system's design.

PROVISION ALTERNATIVES

Aside from "point-to-point" delivery systems, by which products move directly from a vendor to the consumer department, four standard distribution methodologies are commonly used in hospitals. They are: the requisition system, the par level transfer system, the exchange cart system, and the case/procedural cart system.

Requisition System

One of the oldest methods of supply distribution is the requisition and delivery system, sometimes referred to as "fetch and carry." Under this arrangement, someone on the critical care unit has the responsibility for maintaining an adequate level of supplies, for making out a requisition for all needed items, and for physically taking the requisition to central stores for processing. Central stores then fills the order and delivers the requested items on a supply cart to the destination for nursing personnel to make proper disposition. This entire process may be repeated at regular intervals several times a week.

The advantage of this method of distribution is its simplicity. There are, however, many disadvantages. Taking inventory before each requisition is time-consuming. Nursing personnel must take time from patient care activities to handle supply chores, and the salaries of these individuals are generally somewhat higher than the salaries of those involved in the supply process. Furthermore, the entire burden of supply forecasting is placed on the nursing personnel in the critical care unit. This method often results in expensive duplication of inventory, which takes up valuable space. Despite its disadvantages, this method is still frequently used.

Par Level Transfer System

Another system of supply and distribution is commonly referred to as the par level transfer system. Using this method, each department continues to use the same storage space that it did when ordering from central stores. A quota for each supply item is established, and this quota is maintained by central stores. A technician from central stores generally makes the rounds of the critical care units at frequent intervals, approximately two or three times per week and replenishes stock back to "par" levels.

One redeeming feature of this system is that nursing personnel are not involved on a day-to-day basis with supply duties. Unfortunately, the replenishment process is extremely slow with this method because fluctuations in use often make the replenishment process very different each day for each critical care unit and preplanned amounts to restock cannot be planned. With the par level transfer system the hospital still must maintain costly storage cabinets, and costly additional space in each department.

Exchange Carts

The supply exchange cart system works best when carts include all items used by the critical care unit or other departments, from paper clips to intravenous solutions. The carts include a 24-hour complement of supplies for each unit. Duplicate carts are staged in central stores and are exchanged with the carts on the units at predetermined intervals. Supply quotas are established in conjunction with each department, and, to be effective, these quotas are reviewed and updated at least quarterly. Practicality and credibility on the part of the user are of utmost importance with this supply system.

The total supply cart exchange system has proven to be one of the best methods of supply distribution for hospitals regardless of a hospital's characteristics, because it is practical, flexible, dependable, and uncomplicated. It has been established that it saves time, paperwork, steps, personnel, inventory, and money.

The total supply cart exchange system serves five integral functions: forecasting, distribution, supply, control, and patient charge accountability. Foremost among these is the function of forecasting. Because a supply shortage can be anticipated 24 hours in advance of the time the item is needed, the materiel manager has adequate time to prepare and take the necessary steps to procure or substitute for an item. Accurate forecasting is imperative if the total supply exchange cart system is to work. In order to ensure accurate forecasting, all other storage space on the nursing units, such as cabinets, drawers, and storage closets, must be eliminated (7).

Case/Procedural Cart Systems

Case/procedural carts are a highly specialized variation of the total supply exchange cart system. However, as the terms imply, this system is dedicated to highly specialized procedures. Included among these carts are Code Blue carts, dialysis carts, neurological carts, trauma carts, isolation carts, arterial/central line carts, orthopaedic carts, and other carts as the specific critical care unit requires. The carts provide the materiel required for specialty and, often, high-risk procedures in a systematic, controlled, efficient, and cost-effective manner. The carts typically provide the complete complement of materiel for a specific procedure, including the special requirements for various physicians. Included among the distinct advantages of case/procedural cart systems are:

- Controlled, systematic replenishment methodologies;
- Rapid distribution and turnaround time;
- Effective patient charge capturing;
- Standardization of cart set-up;
- Centralized responsibility and accountability within one department.

Practicality and operational constraints usually require the utilization of the spectrum of the provision alternatives discussed.

CONCLUSION

Successfully integrating material services support systems with the daily operations of the critical care units is dependent on all levels of personnel from the nursing department and the materiel services department working together synergistically. The outcome of this work is the right supply at the right place and time, in the right quantities, in the right condition, at the right cost. Effective and efficient materiel support systems are imperative for achieving the highest quality of patient care.

REFERENCES

1. Pirsig RM. Zen and the Art of Motorcycle Maintenance: An Inquiry Into Values. New York: William Morrow & Company, Inc., 1974: p 388.
2. Sanderson ED. Effective Hospital Materiel Management. Rockville: Aspen Systems Corporation, 1985.
3. National Commission on Nursing. Summary Report and Recommendations. Chicago: The Hospital Research and Educational Trust, 1983: p 9.
4. Sanford SJ, Disch JM. American Association of Critical-Care Nurses: Standards for Nursing Care of the Critically Ill. Norwalk: Appleton & Lange, 1989.
5. Seiler WJ, Baudhuin ES, Schuelke L. Communication In Business and Professional Organizations. Menlo Park, CA: Addison-Wesley, 1982: p 4.
6. Crosby PB. Quality Without Tears: The Art of Hassle-Free Management. New York: McGraw-Hill, 1983: p 220.
7. Housley CE. Hospital Materiel Management. Rockville: Aspen Systems Corporation, 1978: p 164–167.

Chapter 9

Computer Technology

JUDITH ANN BLAUFUSS

The first computer was installed in a hospital in the late 1950s and, by 1970, there were more than a thousand computer systems in the health care industry. Computer services now exist in every hospital in the United States to some degree (1). Computers are available to assist staff nurses and nurse managers in most aspects of their jobs; they create challenges and opportunities.

As critical care units developed in the mid 1960s, computers were used to acquire physiological data such as blood pressure and heart rate. Then, during the evolution of hospital information systems (HIS), the patient information was centralized in a patient data file. This phenomenon was of great significance for the critical care professional (2). The ability to communicate and to integrate information, generate records, and assist in clinical decision-making can facilitate appropriate medical and nursing care for critically ill patients and has the potential of having an impact on the quality of "care" as well as its "cost." Both "care" and "cost" are the major concerns of the health care industry and are the driving forces that affect the delivery of care (2–6).

CRITICAL CARE

Gardner et al. (7) outlined how computers used in the critical care settings served four distinct functions in enhancing patient care.

"First, physiological monitors containing microcomputers that acquire, process, store, and display data can sound alarms when important variables become life-threatening. Second, computers facilitate timely and accurate communication of data to and from multiple sources (laboratory, blood bank, surgery, x-ray) within the hospital, making patient information more readily available for use in patient care. Third, medical record-keeping, which enables consistent and continuing patient care, can be enhanced by computers. Fourth, expert computer systems are used to make nursing and medical decisions and to augment the capabilities of nurses and physicians caring for the critically ill" (7,8).

The critical care setting is the most information-intensive area within a hospital. Physicians and nurses collect a large amount of data through frequent observations, testing, and continuous monitoring equipment. As a result, large volumes of data must be stored, processed, and used by the nurses and physicians for clinical decision-making. The computer has become a necessity in critical care because of the large volume of data generated and the limited amount of time that professionals have to respond to life-threatening situations (2,7–9).

The Commission on Nursing (10) recommended the development and use of automated information systems as a means of better supporting nurses and other health professionals. The Commission's decision is supported by the fact that "computers have been used in other service industries to perform well-defined, repetitive tasks and to free valuable human labor for high-level tasks that require the ability to interpret, integrate, and interpolate" (10). With the nursing shortage, any reduction in time spent in recording, tracking, retrieving, and communicating information can reduce demands on nurses' time and enhance an organization's nursing resources.

Nurses compose the largest group of health professionals providing direct service to patients on a daily basis. No other profession has more involvement with hospital information systems that touch all aspects of patient care than does nursing. Nursing's challenge is to use computer technology to innovate and to transform the manner in which nursing care is delivered (1,4). As Hardin and Skiba (11) noted:

"Our decisions must be whether to act as we have traditionally acted and have change thrust upon us from outside the ranks of our own profession, or to anticipate this revolution in our practice, familiarize ourselves with it, and prepare to take an active part in the introduction of computers into the nursing world" (11).

The decisions made today about computer applications in critical care have a direct impact on the future of nursing practice. Nurses need to steer technology in the right direction and, in the process, revolutionize how nursing care is delivered.

COMPUTERS

A computer is an electronic device used to process data. Although there are many ways to define a computer system, it is probably easiest to understand according to the pattern of data flow through the system (1,11–13).

Hardware

A computer system typically consists of the following hardware components:

1. Input device;
2. Central processing unit (CPU);
3. Main memory;
4. Mass data storage device;
5. Output device.

Figure 9.1 shows a sample configuration of computer hardware. Computer systems may have more than one of each of the identified hardware components.

A terminal is an input device which accepts instructions and data from the user, translating them into machine commands. Examples of input devices include:

CRT terminals/keyboards;
Teletypewriters;
Portable terminals;
Hand-held terminals;
Punch card readers;
Optical character recognition devices;
Magnetic ink character recognition devices;
Magnetic media readers;
Voice/image recognition devices.

The CPU is the brain of the computer. It contains the control unit and the arithmetic and logic unit. The control unit supervises and monitors all computer system activities. The arithmetic/logic unit performs all operations directed by the control unit, all arithmetic calculations, and the logical operations and comparisons.

The main memory is logically organized and arranged so that computer instructions and stored data are easily located and accessed. In order to run software/programs, the data and instructions are stored within the main memory. There are two basic types of memory: random access memory (RAM) and read only memory (ROM). RAM is available to the user for storage of data and computer instructions. It is volatile; its contents are destroyed when the computer is turned off. ROM is not accessible for storage by the user; the content of ROM memory is burned into the ROM by the manufacturer, and the user has no control over the content of this memory. The user and/or the software in use can thus read the content of the ROM, but not alter its content.

A mass data storage device provides large amounts of nonvolatile memory that can be used, erased, and reused under control of the user or under software control. It may use a hard disk, magnetic tape, an optical disk, or some other storage media to hold stored information. Access to the data in a mass storage device is slow, and, for efficiency, information stored by a mass storage device is placed in main memory before it is used by the CPU.

The output device or unit provides information to the user. It may also transmit data to a storage media or to other computers. Hardware selection is based on meeting the system performance criteria outlined in Table 9.1.

Software

The other essential component of any computer system is the software that directs the operation of the hardware in its processing of the data. Software is the interface between the user and the computer (Fig. 9.2). Software is often classified into two major types: system programs and application programs.

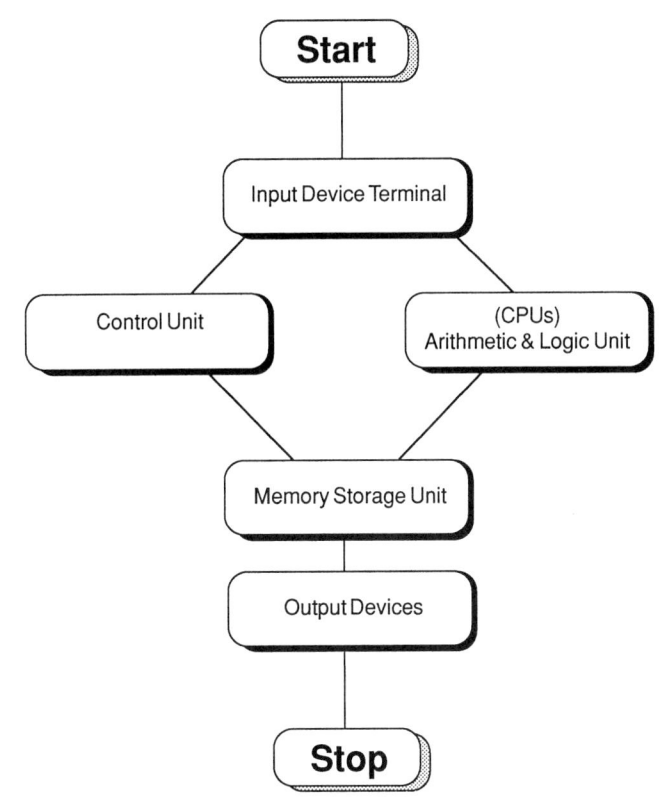

Figure 9.1. Configuration of computer hardware.

Types of Computers

Mainframe computers are large computers. They are typically installed in specially cooled rooms and handle massive quantities of data processing. Minicomputers are smaller versions of mainframe computers. They have the same basic components of mainframe computers, but have been scaled down. Minicomputers are less expensive than mainframe computers, and are generally used for smaller applications. Microcomputers are also known as personal, desktop, or home computers. Microcomputers function similarly to mainframes and minicomputers. All of their components, however, reside as a functional unit capable of fitting on a desk.

The microcomputer is slower and less powerful than the mainframe computer, and the microcomputer is slower and less powerful than the minicomputer. Improvements in computer technology, however, are increasing speed and power at all levels, and high-end microcomputers of today are more powerful than the minicomputers of just a few years ago.

Computer Resource Sharing

Computer resources can be shared through multitasking, multiuser, or network systems. Multitasking refers to a system in which the CPU splits its time among various tasks requested by a user. A multiuser system allows

Table 9.1. Hardware Requirements

---
1. Sufficent processing power to handle the calculated workload
2. Response time of no greater than 2 seconds during peak usage for data entry and retrieval
3. Sufficent online and off-memory for long-term and short-term storage
4. Designed to provide enough data entry ports conveniently placed to permit rapid accessibility to all users
---

Figure 9.2. User computer interface.

several input/output device combinations to time-share a single CPU, and thus the CPU can accommodate several users simultaneously. Local area networks connect multiple computer systems. Rather than consolidating all data in one data base, the network approach allows relevant working data to be stored in each functional module while still providing access to the small percentage of data that actually needs to flow among systems.

NURSING INFORMATION SYSTEM

"Most nurses might not recognize that the first step toward automation is not to address hardware or software requirements but to define the information requirements" (1). If nursing is to benefit from computer technology, nurses in the clinical areas need to define what information requirements should be automated. Nursing also needs to articulate what is the cost-benefit ratio of computerization. Until recently, only pharmacies or laboratories have been able to demonstrate a cost savings (14). As Walker and Schwartz (1) pointed out, it is not as easy to identify the cost-benefits of nursing interventions. They gave the example of the difficulty of quantifying "turning" a patient to prevent bed sores (1).

Ozbolt (15) defined two factors contributing to the delay in the development of decision support systems for the nursing process. The first is that of knowledge representation in nursing. What are the key concepts and issues with which nursing is concerned, how are the concepts and issues related, and how should nursing express them? Nurses as a profession need to agree on standardized language to describe what nurses do (15). Work is occurring in three areas of standardization: nursing minimum data set (16), nursing diagnosis (17), and functional health patterns. (18) Standardization is necessary for automation and comparability within and among health care organizations.

The second factor contributing to the delay in the developing of decision support systems for the nursing process is the lack of knowledge about how nurses make decisions (15). Wessling (19) has observed that more often than not nurses are guided by intuition and subjective thinking that brings inconsistency to their data-gathering process. Wessling feels that "working with a computer forces the nurse to define and analyze nursing and to delineate functions logically" (19). Blaufuss (20) confirmed similar findings during implementation of computerized bedside charting. Computerization caused nursing to develop a logical data base that facilitated charting of the nursing process (20). Once implemented, the system improved documentation of the nursing process significantly when compared with preimplementation studies (20,21).

Steering Committee

It is helpful for the hospital to establish a steering committee including representation from all departments in order to outline a timeline for choosing and implementing a hospital information system (HIS) and identifying and selecting enhancements once the system is implemented. Formation of a subcommittee of the HIS called the nursing information system steering committee (NISSC) has been demonstrated in the literature as a necessary beginning for nursing professionals to organize their needs and strategies (1,3,22). As members of this steering committee, nursing's responsibility is to define its information requirements. Reider and Houser (14) have provided some help by identifying the following six steps for establishing nursing information requirements:

1. Analyze existing forms and reports. The purpose of this analysis is to look at the scope of the information nurses need. For example, to write a patient's care plan, each step of the progress would be examined, including patient assessment, admis-

sion interview, nursing diagnosis, problem list, care plan, medication record, nurses' notes, discharge plan, and so forth.
2. Analyze the information flow and how information is processed.
3. Look at how the gathering of information can be automated.
4. Define the input necessary to generate the required output.
5. Define output information and its format.
6. Define the data in terms of the elements needed for all input and output (14).

The key to having a successful nursing information system (NIS) is to have all aspects of documentation of the nursing process included.

Definitions of NISs vary depending on whether a system is interpreted according to its purpose or according to what it encompasses. The Study Group on Nursing Information Systems describes a NIS as a system that "applies to the automated processing of the data needed to plan, give, evaluate, and document patient care, as well as to collect the data necessary to support the delivery of nursing care, such as staffing and cost" (23).

Saba and McCormick (23) define a NIS as "a computer system that collects, stores, processes, retrieves, displays, and communicates timely information needed to do the following:

1. Administer the nursing services and resources in a health care facility.
2. Manage standardized patient care information for the delivery of nursing care.
3. Link the resources and the educational applications to nursing practice (23).

Four major areas encompassed are nursing administration, nursing practice, nursing research, and nursing education.

Grobe (22) emphasized that a "true nursing system should provide more than relief from repetitive, time-consuming clerical and transcription responsibilities. Genuine nursing systems furnish automated support for nursing planning and patient care. A nursing system should maintain a database to serve as the focus for monitoring, delivering, managing, and evaluating quality nursing care" (22).

The information system must be interactive, providing capabilities for nurses to communicate or talk to other health care areas. If it is set up as a separate computer system, nurses may still be forced into the same role as in a manual system—pulling all of the information areas together and trying to make them relate to each other. If a stand-alone approach is adopted, the nurses' workload for information handling may actually increase, and so will the cost. An example of this is a computer system which is a stand-alone nursing care plan program (NCP). The nurse cannot bring up the final NCP problems on a continuum of care data base program that may include discharge plan. Someone must enter the data from the final NCP into the discharge program. An integrated system does this automatically, eliminating the need for duplication of documentation. Figure 9.3 diagrams the HELP computerized intensive care unit data collection and decision-making system. There is a wide variety of data sources, and data from these sources are integrated into one patient data base. As the data are collected, they are analyzed by the decision-making processor (24).

The NISSC membership list should include a representative of nursing administration. This representative's responsibilities include: (a) representing nursing's information needs to the HIS committee and (b) ensuring that nursing's philosophy and mission are represented in the direction taken by the HIS and NIS. This individual usually becomes the chairperson for this committee. Other members include representatives from all specialty areas within nursing. The majority of hospitals that have implemented either stand-alone or mainframe computer systems to produce reports documenting nursing practice have found it necessary to name a single nurse, or a group of nurses, as computer specialists. These nurses become systems content designers and serve as liaisons between the nursing areas they represent and the computer companies, program analysts, and systems analysts. In some settings, these nurses have received on-the-job training from computer companies and have entered into the more technical end-of-system design and programming (25).

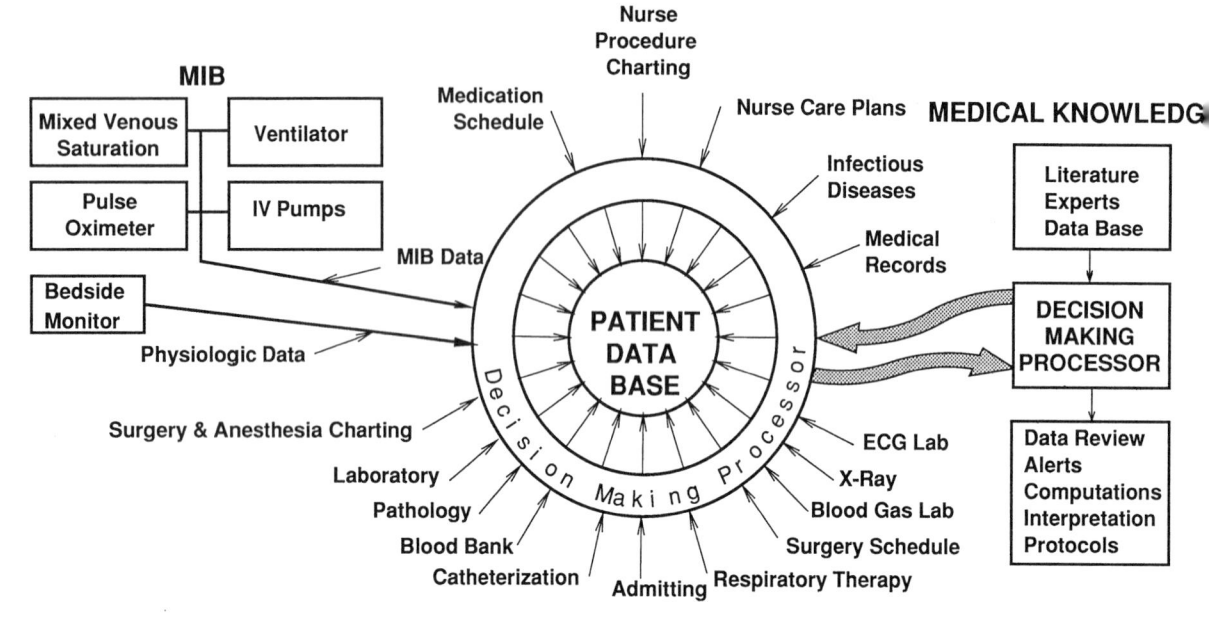

Figure 9.3. HELP computer system. Diagram of computerized intensive care unit data collection and decision-making system. Note wide variety of data sources and how data from these sources are integrated into one patient data base. Also illustrated schematically in the diagram is a medical decision-support system. As data are collected, they are analyzed by the decision-making processor, illustrated by ring surrounding patient data base. (Reprinted from J Cardiovasc Nursing, 4(1):70, 1989, with permission of Aspen Publishers, Inc. Copyright November, 1989.)

Representation on the NISSC should include an administrative representative of the data processing department or biophysics department. This individual's role is to clarify questions concerning the capability of the computer system and to assist nursing in learning to translate documentation needs into a format that programmers can use to develop the software for the NIS if the hospital is a developmental site. If the hospital is not a developmental site, this representative helps evaluate software for compatibility issues.

The NISSC's main purpose is to oversee how nursing prioritizes information needs. NISSC objectives might include:

1. To provide automated support to patient care, unit management, and nursing administration;
2. To improve timeliness, accuracy, and completeness of nursing documentation;
3. To increase efficiency and effectiveness of nurse decision-making;
4. To document and improve quality of nursing care;
5. To identify areas in which to do research in order to identify cost/benefit ratio in relationship to patient outcome (21).

The major responsibilities of the nursing representatives to the NISSC include:

1. Coordinate flow of information from nursing to programmers;
2. Translate nursing needs/requests to software needs;
3. Screen and report design;
4. Test software in laboratory or bench;
5. Assist in training and implementation;
6. Reassess programs continually to evaluate need for changes/updating or maintenance;
7. Analyze problems—hardware, software, and other issues.

User Groups

A user group is a mechanism to involve the staff nurse. Responsibilities of staff nurses who sit on a user group committee include representing their unit's concerns and clarifying objectives of the NISSC. These individuals are involved in identifying the flow of information and how information is processed in order to determine how the gathering of information can be automated. They are often the individuals who begin the process of screen design to report formatting. The user group members are involved in bench testing the software to determine its workability at the unit level. Often user group representatives are able to troubleshoot software before implementation, thus avoiding a breakdown in the computer system leading to frustration of the staff nurses and increasing resistance of acceptance of the computer system (4,22). These representatives act as resource personnel for their units during the training period. They are able to give immediate support and reassurance to their peers, thus establishing the credible link to the NISSC.

For the introduction of a NIS to be successful, Ball et al. (4) outlined three components that must be addressed. They include:

1. Management of the change required;
2. Implementation of the system;
3. Training of the users of the system.

Management of the Change

The introduction of a NIS requires that attention be paid to the hospital environment and to the human factors. Successful management of change results in user acceptance of the system. Poorly managed change generates resistance, and the benefits of computerization are not felt (4,26,27).

Nursing's acceptance of a computer system can be facilitated by preparing nursing personnel for automation and soliciting their participation in the implementation process. This enables them to cope with the changes in their work environment, to adapt to procedural changes, and to utilize the new tool with a sense of some control (4).

Warnock-Matherson and Plummer (28) outlined how nursing managers can promote nursing user acceptance by:

1. Ensuring adequate preparation of the nursing users;
2. Encouraging involvement of nurses in system testing;
3. Soliciting feedback from nurses and responding to concerns;
4. Communicating with nursing staff during implementation;
5. Providing support during the implementation period.

They go on to emphasize the important role that two-way communication plays throughout the entire planning, implementing, and testing periods. User-participation in the decision-making process enables the committee to influence development according to its needs and priorities (28).

Implementation of the System

The key to a successful implementation of a NIS rests on whether or not all the affected individuals and organizational departments have been identified and involved in the planning process (27). Again, the user plays an integral role in identifying possible roadblocks that may cause costly delays. A master plan for the implementation needs to be set up outlining the following:

1. Scheduling of resources;
2. Dividing roles and responsibilities clearly to promote an effective and productive work environment;
3. Setting priorities in relationship to a timeline;
4. Avoiding costly delays and duplication of effort;
5. Identifying personnel requirements;
6. Creating an evaluation tool to identify how to improve on the implementation process.

The term "system conversion" means that a change is taking place from one method of collecting, storing, organizing, and retrieving data to another method. This may be a change from a manual to computerized method, or a change from one computer system to another. There are four basic approaches to system conversion. They include the following:

1. *Direct or Crash Conversion*—this occurs when there is a changeover to a new system all at once. The old system is stopped and is immediately replaced by the new system. The change is abrupt and can cause increased stress to the user. If unsuccessful in its first attempt, it may be very costly in terms of lost credibility.
2. *Parallel Conversion*—both systems are run

side-by-side; or in parallel, for a given period of time. It minimizes failure but is very confusing to the users and increases the user workload for a time, which may be doubled. Provisions should be made to have additional staff available during the conversion period if this approach is used.
3. *Phased or Modular Conversion*—sections of the system are converted and made available to the users at discreet intervals. The advantage is that failure within the program is confined to one area. Training and implementation can be flexible if failure occurs. A major disadvantage is that the new system can be used for specific functions only and that other functions must be performed using the existing procedures. It also takes longer.
4. *Pilot Conversion*—the system is made available to a select unit. Programs are used and evaluated for effectiveness and problem identification, and debugging occurs before other units are brought up. At given time intervals, the remaining areas are given access to the system. The main disadvantage of this type of conversion is the inability to move staff freely between organizational units. Communication problems can occur when pilot group functions must interact within groups not on the new system. It also takes longer.

Combinations of the four types of conversion methods may also be used. As indicated, each method has inherent problems. The goal is to minimize those problems by giving careful consideration to the effect of the conversion method on the organization and to identify a plan to address potential problems. Developing strategies to assist users when they encounter problems can help lessen frustration and enhance users' willingness to work with the system (4,21,29).

Training of the Users

In reviewing current research literature on nurses' attitudes to computers, Warnock-Matherson and Plummer (28) found that nurses, as a group, appear to be resistant to computerization. The degree of resistance varies in relationship with education, length of employment, and work experience. They point out that the attitudes of the nursing staff members need to be assessed before implementation occurs. They suggest that target groups could be identified that require more intensive orientation and education. Additional training and support could be allocated so that the user has a more positive experience (28).

They emphasized that "preparing the nursing staff for automation is crucial in establishing staff user acceptance and minimizing resistance" (28). A training program needs to include:

1. Basic principles of the computer system;
2. Benefits the staff can expect from the system;
3. Roles and responsibilities of the nursing personnel during and after implementation;
4. Communication channels;
5. Orientation of the staff to the system with one-on-one training sessions.

"The attitude of nursing personnel toward the system and, ultimately, their morale, are often a function of the quality of the training provided. Many systems professionals attribute the success or failure of the system to the degree of user acceptance achieved" (3).

The NISSC and user group representatives need to develop objectives for the training sessions. It is essential that this step occurs, not only for specifying the information that is to be presented, but to determine how the evaluation of the training process occurs. Zielstorff (29) suggests that "each objective should be clearly stated, limited in scope, and measurable so that the learner can gauge the degree to which he or she has assimilated the material, and so that the planners of the program can evaluate its effectiveness."

The training session gives an opportunity for the instructor to acknowledge the difference in users' knowledge, attitudes, and skills. By addressing these differences and allowing the learner to express and interact in a non-threatening environment, needed credibility and acceptance for the computer system are gained. The user becomes a willing participant in learning this new technology (27). Implementation may proceed according to the timetable shown in Table 9.2.

Table 9.2. Implementation Timetable

A. Training Schedule
 1. Training as close to time of implementation as possible, preferably a few days beforehand
 2. Decide who will be taught and how many; factors to consider include:
    a. Who will be the most frequent user of the system
    b. Inservice budget
    c. Availability of training facilities
 3. Schedule classes at convenient times, i.e., just before the start of a shift
B. Implementation Schedule
 1. Depending on the complexity of programs, implement a new nursing unit every 2 or 3 weeks; the number of resource personnel available will also affect the implementation schedule
C. On-Call Schedule
 1. Schedule resource persons to be on-call 24 hours a day during implementation period
 2. Develop a mechanism for staff nurses to provide feedback, i.e., communication pad; respond to queries in a timely fashion

Postimplementation Evaluation

This is an extremely important phase. If all is going well, this final phase helps to fine-tune the system that was implemented. If problems exist, evaluation helps to avoid problems in implementing additional applications in the future. Either way, evaluation provides needed insights. Ball et al. (4) pointed out the following questions that need to be asked in the evaluation phase:

1. What reports are being printed and why? Is there a functional use for them?
2. What problem areas are procedural? Who are the problem users and why? Is it because they do not understand the system? Do they need additional training? Do their problems result from not being involved in the implementation process?
3. Is the system doing what the specifications say it is supposed to do? Is the workflow smoother? Is communication improved? Is there any increased efficiency in staff (4)?

Additional questions that the implementation team needs to consider include whether or not the budget and timetable estimates for implementation were realistic. If not, what were the causes of missed deadlines and budget overruns? What obstacles were encountered that had not been planned on? Finally, what is the user satisfaction level? This is of primary importance because if the users are dissatisfied, they will not use the system. Maintenance and continuous attention to user issues and concerns must be a priority if nursing information systems are to be valued (4,22).

NURSING MANAGEMENT SYSTEM

Administrative decision-making is best supported by an automated system that provides timely data from integrated information systems concerning operations, clients, employees, and services. Grobe (22) described the following as essential components of an automated, hospital-based nursing management system:

1. Assisted computerized scheduling;
2. Allocation and distribution of staffing based on patient classifications;
3. Patient data management;
4. Infection control surveillance;
5. Quality control systems;
6. Productivity reporting system;
7. Risk management reporting;
8. Time and attendance record-keeping;
9. Employee data management, including continuing education;
10. Contractual services utilization; and
11. Budget process (22).

Nursing Staffing and Scheduling Systems

One component of a comprehensive nursing management system is the nurse staffing and scheduling process. Staffing systems attempt to qualify several variables as a basis for nurse scheduling. Two major variables include (a) patient criteria, e.g., need for care by condition, ordered procedures, age, state of acuity, nursing skill level; and (b) nursing qualifications, e.g., preparation level, skill level, experience. Reports can be generated that monitor patient acuity levels, budget, nurse workload patterns, and department productivity (5,22). Overall, these systems give nursing the control over staffing costs and more effectively relate staffing to patient care requirements.

"Intelligent" systems have the capacity to adjust staff schedules in an interactive fashion on a shift-to-shift basis. They are driven by the workload that the staff nurse generates. The staff scheduler then creates a schedule that incorporates nursing requirements (determined through measurement), the various levels of staff expertise (based on data from the personnel system and education records), contractual specifications, and personnel policies (4).

The benefits of such a system are many. Advance scheduling is helpful to both the manager and the staff. The computer is more able to meet staff individual needs for time off and do it in a far more objective way. Record-keeping is objective and enables the manager to monitor absenteeism, scheduled time off, and turnover (4,5,22,29).

Budget

The next challenge for nursing is to develop patient classification systems that identify nursing department costs and revenues. Such a system has been developed as part of the nursing component of the HELP (Health Evaluation Through Logical Processing) system. It is a care-based system that identifies the actual nursing care delivered to each patient rather than categorizing patients into levels of care. It is an example of an integrated HIS. As nurses document care given to a patient, the computer automatically flags key patient care activities and sends them to an acuity file. At the end of a nursing shift, the time allotted for all the nursing activities in the patient's acuity file are totalled by the computer to determine the nursing care minutes for each patient on that shift. Individual patients are then grouped by nursing unit to determine the nursing hours per unit and staffing needs prioritized by actual patient care needs. Staffing mix and patient acuity are matched and appropriate staff is assigned (5).

QUALITY ASSURANCE

Ability to computerize a quality assurance program that is integrated with the HIS has the potential of not only ensuring that quality of care is provided, but that it has direct measurable impact on dollars saved by the hospital. Such an information system can provide verification and warning of potential errors before errors occur, thus making a major impact on delivery of quality care. The two major areas nationally where quality of care and cost have been studied are in regard to medication errors and patient falls. Any impact computerization can have on decreasing costs related to complications, increased length of stay, and patient safety will be significant (30).

The future quality assurance capabilities of an integrated system may include that nursing could actually reduce staff by reducing the number of quality assurance staff needed as well as freeing the staff nurse from data collecting of patient specific monitors. The computer has the capability to retrieve online concurrent data for nursing audits for monitor purposes, which saves considerable time and staff nurse involvement (4,22,30).

SECURITY

The proliferation of computerized clinical and management information systems in health care has led to concerns related to protecting the privacy of and ensuring the confidentiality of health data (31). Therefore, overall consideration must be given to the development of institutional specific confidentiality and security policies and procedures. Table 9.3 clarifies terms used in regard to security.

The security of information can be achieved by restricting access to the terminals, programs, files, and records. Security of this type is normally based on the user keying in a password that is checked by the computer to determine the level of accessibility permitted for that user. The password system can lock a user out of using specific terminals, programs, and files. For example, a certain password might allow someone to look up, change, and print a copy of a record. Another password level might allow the user to see the record but not to change it. Yet a third level might not allow the user to see the record at all (1,31).

FUTURE

The computers of the future will have radically different software and hardware. They will process large amounts of data faster than presently possible, solve complex problems, make inferences similar to human reasoning, and process natural language.

Table 9.3. Security in Computers

Data Security	Security protects data against deliberate destruction, accidental access, or loss by unauthorized persons
	Data security involves protecting the hardware and software as well as the physical security of the computer system equipment
Physical Security	A computer system requires protection from natural disasters such as fire and floods, and from deliberate, destructive acts
Data Protection	Protection of data during transmission to and from the computer system is also part of data security
	Data protection involves methods to ensure that all data input, processing, and output are accurate
Privacy	Privacy of a computer system ensures people that personal data is protected against improper access

Staff Nurses

The ability to have large data banks of nursing information for which computer analysis has been applied in order to assess nursing actions, independent decision-making, and patient outcomes may identify which combinations of variables are powerful predictors of patient outcomes. Specific predictors of individual or group data on patients can assist the staff nurse in evaluation and decision-making (23,25). McCormick and McQueen (25) have noted that "diagnosis and prognosis based on systematically collected and collated data are more accurate than those based on clinical judgment and intuition alone. Decision-making can become more scientifically based as the probabilities of outcomes associated with alternative means of treatment can be compared" (25). This will give the staff nurse a sense of control and will help to transform the complex health care environment into a more logical structure.

Nurse Managers

With an increased flow of information from the patient level to the administrative level, the manager needs to be aware of statistics, cost data, and have the ability to articulate the cost-benefit ratio of program institutes. The computerized data from the HIS, NIS, and NMIS can be readily retrieved and used as support in decision-making.

Critical Care Units

Gardner et al. (8) asked the question whether or not computers can successfully match the needs of intensive care units (ICU). They believed it revolved around two issues: "First, is there a critical mass of patient information in the computer database and second, has the ICU developed an adequate computer society with the maturity to live through the change in the style of patient care required by computer implementation (8)?" Both issues must be resolved for ICU computer systems to be properly implemented.

Organizations

The focus is shifting from a nursing care approach to a patient care approach. Ball et al. (4) pointed out that there is a paradox because we are in an age of increasing specialization and yet the integrated computer system focuses on a return to a centralization of patient care functions. "Few health care organizations will be able to support five or six seperate and distinct administrative departments to provide one service, patient care" (4). Hospitals will have decentralized management and centralized care delivery. This can only happen with the assistance of an integrated computerized information system.

The author wishes to acknowledge Reed Gardner, PhD, Carol Ashton, PhD, RN, and Debbie Hinson, MS, RN for their assistance in completing this chapter.

REFERENCES

1. Walker MB, Schwartz C. What Every Nurse Should Know About Computers. New York: JB Lippincott Co., 1984.
2. Avila LS, Shabot MM. Keys to the successful implementation of an ICU patient data management system. International Journal of Clinical Monitoring & Computing 1988;5:15–25.

3. Romano CA. Development, implementation, and utilization of a computerized information system for nursing. Nurs Adm Q 1986;10:1–9.
4. Ball MS, Hannah KJ, Jelger UG, Peterson H. Nursing Informatics: Where Caring and Technology Meet. New York: Springer-Verlag, 1988.
5. Budd MC, Propotnik T. A computerized system for staffing, billing and productivity measurement. J Nurs Adm 1989;19:17–23.
6. Randall A. It's time for the next generation of patient care systems. US Healthcare 1989: p 54–56.
7. Gardner RM, Bradshaw KE, Hollingsworth KW. Computerizing the intensive care unit: Current status and future opportunities. J Cardiovasc Nurs 1989;4:68–78.
8. Gardner RM, Sitting DF, Budd MC. Computers in the intensive care unit: match or mismatch? In: Shoemaker WC, Ayres S, Grenvik A, Holbrook PR, Thompson WL (eds). Textbook of Critical Care Medicine, 2nd Ed. Philadelphia: WB Saunders, 1989.
9. Leyerle BJ, LoBue M, Shabot MM. The PDMS as a focal point for distributing patient data. International Journal of Clinical Monitoring & Computing 1988;5:155–161.
10. Secretary's Commission on Nursing Final Report. Washington D.C., Department of Health & Human Services 1988;1:24.
11. Hardin RC, Skiba D. A comparative analysis of computer literacy education for nurses; SCAMC. Silver Spring: IEEE Computer Society Press, 1982: p 525.
12. Brooks FP. The Mythical moon-month, essays on software engineering. Reading, PA: Addison-Wesley Publishing Co., 1982.
13. Frenzel LE. Crash Course in Microcomputers, 2nd Ed. Indianapolis: Howard W Sams & Co., 1984.
14. Reider K, Houser M. Identifying requirements for a nursing system; SCAMC. Silver Spring: IEEE Computer Society Press, 1983: p 475.
15. Ozbolt JG. Developing decision support systems for nursing. Comput Nurs 1987;5:105–111.
16. Werley HH, Lang NM, Westlake. Brief summary of the nursing minimum data set conference. Nurs Manage 1986;17:42–45.
17. McLane AM (ed). Classification of Nursing Diagnoses: Proceedings of the Seventh Conference. North American Nursing Diagnosis Association. St. Louis: CV Mosby, 1987.
18. Gordon M. Nursing Diagnosis: Process and Application, 2nd ed. New York: McGraw-Hill, 1987.
19. Wessling E. Automating the nursing history and care plan. J Nurs Adm 1972;2:34–38.
20. Blaufuss JA. Promoting the nursing process through computerization. In Salamon R, Blum B, Jorgensen M (eds). MEDINFO 86 BV North-Holland: Elsevier Science, 1986: p 585–586.
21. Johnson DS, Burkes M, Sittig D, Hinson D, Pryor TA. Evaluation of the effects of computerized nurse charting. In Stead WW (ed). The Proceedings of the Eleventh Annual Symposium on Computer Applications in Medical Care. Silver Spring: Computer Society Press, 1987: p 363–367.
22. Grobe SJ. Computer Primer & Resource Guide for Nurses. Philadelphia: JB Lippincott, 1984.
23. Saba VK, McCormick KA. Essentials of Computers for Nurses. Philadelphia: JB Lippincott, 1986.
24. Gardner RM, Bradshaw KE, Hollingsworth KW. Computerizing the intensive care unit: Current status and future opportunities. J Cardiovasc Nurs 1989;4:68–78.
25. McCormick K, McQueen L. New computer technology. In: Series on Nursing Administration. Menlo Park, CA: Addison-Wesley Publishing, 1988: p 58–69.
26. Kirkpatrick DL. How to Manage Change Effectively. San Francisco: Jossey-Bass, 1985.
27. Hodgdon JO. ADP management problems and implementation strategies relating to user resistance to change. In Dunn RA (ed). Proceedings of the Third Annual Symposium on Computer Applications in Medical Care. New York, IEEE Computer Society, 1979: p 843–849.
28. Warnock-Matherson A, Plummer C. Introducing nursing information systems in the clinical setting. In: Ball MJ, Hannah KJ, Jelger V, Peterson H (eds). Nursing Informatics: Where Caring and Technology Meet. New York: Springer-Verlag, 1988.
29. Zielstorff RD. Computers in Nursing Administration. In: Heffernan HG (ed). Proceedings of the Fifth Annual Symposium on Computer Application in Medical Care. New York: IEEE Computer Society, 1981: p 717–721.
30. Blaufuss J, Tinger A. Computerized falls alert—a new solution to an old problem. Proceedings of Twelfth Annual SCAMC Conference. Silver Spring: IEEE Computer Society Press, 1988: p 69–71.
31. Romano CA. Privacy, confidentiality, and security of computerized systems—The nursing responsibility. Comput Nurs 1987;5:99–104.

Part V

PHYSICAL SPHERE

Chapter 10
Physical Plant Design and Equipment Procurement

DIANE E. NITTA, GEORGE HICKEY

Planning and designing the critical care environment must result in "user friendliness." Nurses are the primary users of well-planned and smoothly operating care areas, and patients are the recipients. The critical care unit is a microcosm of the hospital environment, whose functional requirements are integral to the needs of nurses, patients, physicians, other hospital staff, and materiel distribution systems (1).

In order to provide the optimal design for an intensive care unit space, nurses must be key members of the planning team. The ability to contribute to an effective patient care unit design, however, requires the nurse to have an architectural design perspective as well as a nursing practice perspective (2). The nurse translates family needs, patient care needs, and operational needs to the architect.

The planning process for critical care unit design is dynamic and must remain fluid for the duration of the project. Depending on the completion time, many factors may change, e.g., personnel, funding, and completion schedule dates.

PLANNING

The planning team convenes early in the conceptual stages of the project, even before selecting the architect or establishing the working budget. Effective planning teams are comprised of core representatives from appropriate subcommittees, which are formed as project needs dictate. Key members of the planning team include representatives from nursing, administration, medical staff, and ancillary and support services. A project coordinator is usually appointed.

Goals of the planning team include planning for realistic space and program needs, not just the maximum number of beds. The planning team provides clinical and operational input into the design of the unit and identifies equipment needs.

Two important subcommittees of the planning team are:

Project Policy and Resource Subcommittee—to plan strategies for accomplishing construction and securing equipment, staging the project, and tracing the budget. Members include the architect, the administration, the project coordinator, and a clinical engineer.

Planning Subcommittee—to ensure appropriate integration of all planning activities and reach agreement on clinical or operational issues that affect more than one individual unit. Members include staff nurses, the critical care nurse manager, a physician, a hospital administration representative, and the project coordinator.

The planning team's existence encompasses the entire duration of the project, from first conception to project opening and beyond, and includes follow-up on unfinished or unsatisfactory aspects of the project. The goal is to create and produce an efficient, effective facility for the future as well as for the present.

Second only in importance to the planning team is the ability to create a full-size mock-up space wherein aspects of room and central control station layouts can be explored in conjunction with the usual and emergency equipment, e.g., nuclear cameras, portable x-ray machines, hypothermia machines, multiple IV pumps, etc. The mock-up space will have a direct and positive impact on facility design and give the planning team more authority when working with the project architect. If possible, the mock-up space should be kept intact throughout the completion of the project to address decision points when required by the architect.

Planning the interior design, which includes wall and floor coverings, color schemes, furniture, and seating arrangements, should be done by utilizing an expert in health care environments. Pragmatic critique of interior design selections should be provided by the planning team, especially nursing members.

PROJECT EXECUTION

Projects classically follow the same development process: (a) conception of need or opportunity; (b) definition of design; (c) architectural design and creation of construction documents; (d) selection of contractors and suppliers; (e) construction, including change orders during construction; (f) beneficial occupancy and move-in logistics; and (g) making the project work.

Each project step is critical, defining a clear, definite historic event. Even though sequential and discrete, the project pathway is still long and tortuous with steps often overlapping. One role of the planning team is to keep the project process moving while identifying potential problems and solutions. To be most effective, the pathway should be understood by all involved.

Conception of Need

When first recognized, need for change is usually difficult to define. To move from an informal, casual idea to a well-established project, the need must be demonstrated. Working with administration, the planning team should define the need, delineate alternative solutions, identify drawbacks and advantages for each solution, identify perceived costs and savings that are reasonable to estimate, and incorporate best dollar estimate of intangible costs and savings. The objective is to draw up a list of costs and savings for each alternative much like a balance sheet.

The importance of the nurse manager and staff members devoting time to analyzing possible projects is that practical operations are understood in detail. Nursing staff members working with the planning team can add pragmatic parameters to design guidance from the beginning.

When the need for change is accepted, the project becomes "feasible." The challenge for nursing staff members is to assist in making the project practical. When an "input" opportunity occurs, nurses must be prepared, direct, and clear.

Definition of Design

The planning team need not wait for an architect to be identified for the project before beginning the design. The team can proceed with rough sketch layouts of possible patient care unit space using blocks (constructed from paper, cardboard, or any accessible material) as models to represent room units, maintaining spaces for potential corridors used by staff, visitors, and materiel (materiel logistics).

Preliminary work in materiel logistic planning can be of significant help to the architectural effort. Various functional block layouts can be tried by moving horizontal and vertical elements. The planning team can identify and separate staff and visitor ingress, egress, and control, and the flow and storage of clean and contaminated materials. Location and types of isolation, treatment rooms, and other core functions can be approximated. The team can also rough-in the vision span from nurse station locations.

Architectural Design and Creation of Construction Documents

Another product of the planning team is a punch list of functions and physical issues important to the staff for architectural guidance. This type of information is a direct aid to the architectural "value engineering" effort, which is the next process.

The architectural design is in at least two steps: design development and construction documents. If the planning team's efforts are managed well, the guidance criteria given to the architect should make an impact on the physical layout and design. At this point, the architect initiates design development and interacts with the planning team when choices must be made because of physical and cost constraints. Part of this process is to estimate project costs, and the team must insist on comparative alternatives as an element of the decision process.

When the architect feels that the design development phase is near completion, the planning team is asked to sign off on the development plans. The drawings are reviewed carefully, and specific items of concern listed as part of the sign-off.

Upon acceptance of the design development set, the architect creates construction documents. This is an intense period of time. Construction documents consist of a complete set of building plans covering the architectural, structural, mechanical, electrical, and plumbing layouts, as well as the interiors of the project. The documents also include a book of project specifications. The planning team should become familiar with the building plans and the project specification book.

The planning team is asked to review the construction documents at the "50% level." This is the last chance to delineate specific items of concern. Time should be spent with the documents in reviewing each item's disposition, being aware of unannounced changes forced by real or perceived design constraints. Although changes at this time cannot be major, the planning team is able to fine-tune the project after spending time with the specification book and drawings. The architect works assiduously to develop the design presented; most architects are not used to "owners" that are as involved and knowledgeable as the planning team may be during this time.

A sign-off is requested and architects contend that any delay at this point is going to have a serious impact on the project. The hospital, however, lives with the results for many years and the cost-benefit of each change needs to be identified. The 100% construction documents are then presented for review and approval. Change at this point is very unusual, and in fact, should not be suggested by the team unless some unexpected, critical situation exists. The planning team's list of concerns should be maintained for later use. The project now "goes to bid."

Selection of Contractors and Suppliers

The hospital, directed by the architect, uses the construction documents to bid out the job. A general contractor is selected to assemble a group of specialized subcontractors to build the project as designed. The general contractor may have a subcontractor specialty, in which case the general contractor would undertake that particular subcontracting area.

The design and construction process takes significant time. Changes do occur because of regulatory bodies and agencies, conditions that were not apparent during design, and the direction of the hospital. Each of these "change orders" during construction cost money and time and should be avoided if at all possible.

The planning team must try to minimize or even ban change orders. Controlling project changes after the contractor has been selected helps conserve the hospital's funds and gets the project completed and functioning sooner than if "carte blanche" were given for all changes. Experience shows that changes needed by the hospital are lower in cost if they are done after the project is completed, particularly if these changes can be done by hospital personnel or local tradespeople.

Part of the general contractor's bid includes all fixed equipment that makes the structure work. This type of equipment includes vertical transportation, air handlers, chillers, boilers, pumps, and many other devices that are needed for the building to function. Some of the equipment included in the contractor's responsibility are clinical in nature and directly involve staff. These types of devices involve communication systems, surgical and exam lights, patient room headwalls, sterilizers, and other similar pieces of equipment and systems. It is recommended that these types of equipment be removed from the contractor's responsibility for specification and purchase. The contractor may remain responsible for the installation of some of this equipment.

Construction

Fixed, large devices using building utilities, such as steam and ventilation, are candidates for contractor installation. The hospital should retain the right to specify and purchase these devices to assure control over type, function, and cost. The hospital must assume responsibility for timely delivery of devices suitable for rough-ins as called for in the construction documents.

Other systems such as intercoms and premanufactured headwalls can be installed under the responsibility of the hospital. The purchase of these systems by the hospital will include installation responsibility.

Benefcial Occupancy and Move-In Logistics

From the contractor's point of view, at the completion of the project, the hospital is offered beneficial occupancy and the project is signed-off. Before this event, the architect completes a deficiency punch list and the contractor corrects most of the items. The planning team members must assure that they have thoroughly inspected the project and have provided the architect with the team's punch list of items before this time. Some of the items on the architect's punch list are corrected by the contractor after beneficial occupancy. The planning team must know what these residual corrections are and must ensure that the unit does not open if any of these items affect operations.

Upon beneficial occupancy, the ball is back in the planning team's court. The team has been preparing for this moment. During construction, concurrent planning activities by the users must occur. Whether the project is new construction or remodeled space, staff nurses must be prepared to put it into full operation at the time of occupancy. A nurse project coordinator who is part of the planning team can be most effective in coordinating operations with project completion.

Interdepartmental Impact

How will the new unit make an impact on the workload or schedule of support and ancillary services? Communication with all hospital departments is essential to ensuring smooth operation of the unit. It may be helpful to institute a checklist of all departments, managers, and anticipated impacts, in areas such as inventory levels, delivery routes and times, and frequency of replenishing supplies.

New Equipment

Inservices on new equipment can occur during the weeks before opening the new unit. The vendors, nursing education, and nursing staff should be utilized to provide inservice and "hands-on" demonstrations. Purchase terms and conditions for sophisticated equipment should incorporate vendor training responsibility, including nursing control over when and how much training is to be provided.

Space Orientation

Nursing staff members may have been involved at one point or another in the project design and set up of the mock-up space. In the few days before move-in, it is invaluable to have nursing staff members conduct their own environmental orientation in the new unit. This can consist of mock codes, equipment and supplies "scavenger hunts," operation of all equipment, and even moving a gurney or bed through the area to develop a sense of familiarity with the area, as well as to discover what does not work and will need attention before move-in.

Equipment Inventory

Even though the purchasing department may handle the mechanics of equipment purchase and delivery, it behooves the nursing project team member to verify the ordering, delivery, and installation of all moveable equipment.

Making the Project Work

The best planning and preparation does not eliminate the fact that change will occur. How the staff members respond to the change and deal with the pre- and postmove stressors may depend on several factors:

- Quality of teamwork before the move;
- Quality of the relationship between staff members and the nurse managers;
- How well staff members know the clinical specialty of the unit;
- Attitude of staff members—do they believe "fix it or step around it; it's not worth fretting about"(3).

The quality of staff involvement and ownership of the new unit is related to the ability to "roll with the punches" and effectively utilize a feedback loop for addressing residual operational issues.

KEY DESIGN FACTORS

The planning team considers many aspects in developing the project, which produce significant questions and discussions. The nursing members of the team can use the American Association of Critical-Care Nurses (AACN) s for the care environment of the critically ill patient (4) as guidelines.

Patient Care Space

The critical care patient space is often visualized as a private room that has hard, "sound-tight" walls on the back and two sides and that has a front with adequate access. It seems that there is never enough space in a critical care patient room. Often the space itself is adequate but the layout of equipment, facilities, or room proportions create difficult crisis management. A minimum dimensional guideline for a simple room layout is: 4 feet free on each side of the patient bed and 6 feet free and clear lengthwise, 2 at the head and 4 at the foot. More than 5 feet on each side and 8 feet in length becomes difficult for "reach and grab" access to supplies and equipment. It is recognized that there are environments where hard walls are inappropriate or not possible. Although the walls may be curtains in these cases, the dimensional guides should prevail. If equipment pedestals or utility columns are planned, more floor space may be necessary. Also, additional lateral space is needed for such things as supply storage, charting desks, hospital information systems (HIS) terminals, and hand-wash sinks.

The walls to adjoining rooms can be fitted with double-pane vision panels equipped with integral blinds controlled from either side. "Sound-tight" tells the architect to include insulation and ensure that built-in utility boxes (e.g., electrical outlets) should not be back-to-back between rooms. Ceilings, while providing access, should also afford sound control and should contribute to room hygiene.

Room front access should be complete. Before the advent of nuclear cameras, a 4-foot door was adequate. Because of these machines, radiolucent beds, and other devices, "store fronts," sliding, folding, safety glass doors are now recommended. These can be obtained without floor tracks to facilitate equipment maneuvering.

The ability to see the patient from nurse stations is of paramount concern to the staff and regulatory bodies. In laying out a patient care unit, overall patient room width becomes significant because the number of rooms that can be seen from the display area of a nurse station is limited by the room width. Store fronts are a clear advantage to visualization. The planning team may want to use vision lines to analyze unit layout. A guidance criterion is the ability to see the patient's upper body. Closed circuit television (CCTV) is sometimes employed as a substitute for visualization, but regulatory agencies object to this. Signal conduits are often planned for CCTV in each patient room, but for possible future demand, not for visualization.

Lighting

Lighting has been addressed using a variety of techniques in patient spaces. Lights in headwalls, portable lights, surgical lights, ceiling-mounted, motor-aimed high-intensity lights, and light pipes are among the many approaches that are tried for patient illumination. Four uses of lights in a patient space are: general illumination, patient reading, night light, and patient examination. General room illumination that is controllable by staff and patient can be obtained from a headwall laminar fixture with upward-directed light at 50+ foot-candles of illumination. Patient reading lights, controllable by staff and patient, can issue from a headwall laminar with downward-directed light at about 75–100 foot-candles. The night lights, one on each side of the bed near the floor, should be on continuously, about 15 watts each. This allows staff members to enter the room at night, see floor obstructions, and yet not disturb the patient. Patient examination illumination is consistently a major issue. The guideline is to minimize shadowing of color corrected light with an intensity approaching 200 foot-candles at the bed surface. Two simple, effective fluorescent ceiling fixtures each consisting of four tubes, 4 feet in length aligned head to foot with the center of the bed may best meet the general illumination needs. Normally this light is off, but when used, it is controlled by staff members for work with the patient. Hard shadows and heat produced by focused incandescent or high efficiency lamps are virtually eliminated. When detailed work such as a cut down is needed, a portable examination lamp can augment this illumination.

Electrical Power

Techniques and technology become more elaborate through time and constantly add to the electrical energy demand in the critical care environment. The main objective is to provide plenty of reliable power for patient care. Current recommendations are for two independent 20 Ampere (A) circuits per critical care patient room on emergency power for clinical equipment. The receptacles should be located on each side of the bed and the two circuits mixed so that if one circuit is lost there are remaining active receptacles available on both sides of the bed. The next circuit, also 20 A and on emergency power, contains one set of outlets on each of the walls that is not used for the patient's headwall. These outlets are to be reserved for emergency teams responding to crisis management and should not be used for other purposes. A fourth circuit, also on emergency and shared between two rooms, is a "high amperage" 120-volt circuit—30 A—available near the foot of the bed on a side wall. This is used for such high amperage devices as thermia units, air float beds, and other creations possible in the future. The fifth circuit, possibly on "normal" power and shared by several rooms, is used for the patient's entertainment T.V., housekeeping, and miscellaneous purposes. Finally, a sixth circuit, emergency power preferred (but if cost tradeoffs demand, normal power), is a 220-volt 30 A circuit for several rooms, with receptacles usually located in the corridor between rooms. This is used for image intensifiers, cameras, and other mobile devices. It is recommended that receptacles be located well above the floor to eliminate stooping and to encourage proper removal of plugs.

Communications

Communications are several separate entities, one or more of which are in the process of coalescing at any given time. Nurse call, staff page, overhead page, telephone, data, dictation, and physiological monitoring are all communications entities.

Signal systems (conduits, cable trays, and patch boards) should be developed for each, tying the systems together at central points. On some projects, a separate functional space is developed for power, uninterrupted supplies, computer mainframes, and technician work space for each critical care unit.

It is projected that development in this field will evolve into the following: (a) nurse call + nurse page + monitor alarms; (b) telephone + dictation + data; (c) monitors + data; and (d) with investment of personal page, deletion of overhead page. Ultimately, all will merge into one interfaced system.

Heating Ventilation and Air Conditioning

Heating ventilation and air conditioning (HVAC), a separate and distinct architectural and construction discipline, is one that the planning team should address. Although normal considerations for clean, conditioned air is completely handled by the professionals, the designers need to know the total heat load in the space. The planning team should project the equipment that will be used in each space with contingency planning for future equipment. This information will allow the architect to determine the size of the facilities to be provided. A consideration of "worst case" conditions, including the use of hypothermia, should be provided.

All projects include relative air pressures. Important in patient care, this concept means "which way the air blows" and is generally specified by design codes for critical care units. If medical care dictates special isolation techniques, the planning team needs to identify this for the architect. Often, "clean air" rooms (so-called laminar flow) are needed for some patient management; this specific requirement needs identification early in the project design. An example of impact on HVAC design and apparent concern for lighting is the use of switched ceiling-mounted heat lamps (in lieu of radiant heat panels) for burn care units. Once identified, the architectural team will factor in the loads as needed.

Plumbing

The main involvement for the planning team in the subject area of plumbing deals with medical gasses and the issue of interference of other equipment preventing the ready access to medical gas outlets. The planning team should carefully lay out gas outlets in the mock-up rooms to assure clearance for regulators, bottles, and tubing management.

Equipment

The use of mock-up spaces is extremely productive in designing headwall/equipment layouts. The planning team should identify the location and range of movement for each type of equipment projected for use in the environment.

Some important considerations on this planning area are to keep equipment off the floor and to keep all controls accessible. Differences between care units can be significant; cardiac, respiratory, surgical, neonatal, and burn all have widely varying equipment needs. Differences between facilities are also critical (local community hospital versus university teaching hospital). The planning team must clearly identify the mission and the "worst case" scenario of equipment concentration that might be used. This requires projections into the future, including management of equipment not yet available.

The planning team should start with a generic punch list of known and projected equipment. The following lists are provided for guidance:

Assigned Equipment:

1. Physiological monitor;
2. Monitor recorder;
3. Data terminal;
4. IV pumps (worst case quantity);
5. Respirator, volume and IPPB;
6. Respiratory gas monitors;
7. Defibrillator;
8. Regulators—oxygen, air, and vacuum;
9. Elapse timer;
10. Sphygmomanometer;
11. Nurse call/T.V. controls;
12. Code Blue (each side of bed);
13. Transducer mounts;
14. Electrical outlets;
15. Light controls;
16. Telephone/computer jacks;
17. Thermia unit;
18. Intra-aortic balloon pump;
19. Orthopaedic frame;
20. Pumps—feeding, suction;
21. Patient bed;
22. Accessory and supplies storage, breathing bag, airway, cut down tray, pacemaker tray, emergency tray, catheters.

Itinerant Equipment:

1. X-ray, mobile;
2. Image intensifier;
3. Nuclear camera;
4. EEG machine;
5. EMG machine;
6. Crash cart.

Most equipment is located on the patient's headwall. Some is mounted on ceiling tracks. Floor-standing equipment (volume respirator, thermia) must be assigned functional design space. Centralized thermia and defibrillator systems have been implemented in an effort to relieve floor space in critical care. Their use is not widespread, however, because of cost, control, and maintenance considerations. Oval IV track with a minimum of two equipment trees located anywhere at the side, head, or foot of the patient is to carry light loads only, e.g., IV bags and bottles. When more than four IV pumps are to be used, it is recommended that two straight equipment ceiling tracks on the outside of the IV track be added. These tracks should be specified to carry the weight of several IV pumps.

The need for access in order to be able to monitor controls may force reassignment of the monitor to the center or right of the headwall. In an attempt to keep the field clear, the physiological monitor may be fitted with rear entry for patient inputs with the cables running inside the headwall to a junction box near the head of the patient bed. This is dependent on the manufacturer of the system finally selected, since some contemporary systems do not allow back panel patient signal input.

When the term "headwall" is used, it implies either a manufactured, preassembled wall mounted to the building structure or all utilities, including equipment mounts, built into the building wall itself. It is best to bid headwalls out and compare the cost to the contractor's cost of providing the same features installed directly in the building walls. Flexibility is a consideration, but if the planning team is successful with the mock-up layouts, future changes are minimal. It is possible to make changes in headwall utilities even if they are installed in the building walls. If a preassembled, manufactured headwall is selected, it is necessary to ensure that the manufacturer has the responsibility of full assembly and site installation according to the building contractor's rough-in facilities. These responsibilities should be factored into the cost of the manufactured headwalls.

Nursing Station

Nurse space, physician space, and clerical space need to be clearly defined. The nursing station serves as the hub of activity for information flow and processing patient data. Separate cabinetry and writing surfaces equipped with dictation equipment and easy access to patient charts for the physician diminish congestion. Clerical space should be treated as sacred and should not become a place to congregate for anyone spending time in the nursing station. The clerical telephone should be off-limits to anyone other than clerical personnel.

Aspects of nurse station design usually address visualization, layout, maintenance, and heat management. Patient visualization from the nursing station is a major design concern. This is defined by the layout of the patient rooms relative to the nursing station and the type of corridor walls provided for the rooms. Because visualization has limitations, two considerations should be quickly eliminated: elevated nursing stations and "lab bench height" cabinets. The elevated nursing station is raised one or two steps to allow a supposedly greater visual coverage. In analysis, there is little expansion of visual range laterally, and staff fatigue toward the end of a shift makes this approach undesirable. Lab bench height cabinets address the height of the writing surfaces inside the nursing station. Normally, cabinets are a height of 29 or 30 inches off the floor, but the lab height of 36 inches is sometimes considered. This requires higher seating as provided by drafting stools or lab stools. While this appears to encourage movement in and out of seating, lab height is rarely recommended after mock-up trials. One issue is the ease of seeing over the top of the equipment when seated. If the interior writing surface is at normal height, the maximum height of the cabinet enclosing the display equipment should be 44 1/2 inches or less. Anything more than this makes direct visualization of patient rooms increasingly difficult.

The layout of equipment in a nursing station, including monitor displays, computer terminals, and intercoms is a major design effort. Despite the thoroughness of the planning team, relocation of work stations and changes in equipment always occur throughout the lifetime of the facility. Cabinet design can make relocation and changes in equipment very practical. Change can be accommodated if large, accessible cable chases and movable support partitions are designed in the cabinetry from the beginning.

Cabinet design also has an impact on approaches to equipment maintenance. A significant amount of equipment maintenance can take place outside of the nursing station. Impact on nursing function is decreased if removable access panels are provided on the corridor side of the cabinets.

All electrically powered equipment generates heat. When the equipment is contained in cabinets, the heat generated must be properly removed or temperatures build, and working conditions for nursing staff and equipment become undesirable. An ideal solution is to disperse the heat directly into the hospital HVAC exhaust through the corridor toe space. Large area exit vents are developed in the reveals of the upper edge of the removable cabinet panels along the corridor. If the natural convection of the chimney effect needs augmentation, quiet "muffin" fans can be located within the cabinet to enhance the convection effect.

Treatment Rooms

Treatment rooms are defined by the type of function performed. The general purpose treatment room is simple and straightforward with design standards well understood to include adequate space, light, oxygen, vacuum, storage, sink, work surface, intercom, and emergency call system (Code Blue). Adding specialty functions changes room features.

Exercise rooms call for basics, adequate HVAC considerations as well as an area for rest, space for a crash cart, and management of arrests. The size of the room is dependent on the number of patients to be managed and the type of exercise equipment used. When monitoring by telemetry is employed, a desk display is needed, as is a backup display at the nearest cardiac intensive care unit. Besides the normal work surfaces, a nourishment refrigerator and drinking water is desirable.

Endoscopic and surgical exam rooms require additional storage facilities; both often need storage for fiberoptic equipment and catheters. The surgical room needs surgical-type ceiling lights, scrub facilities, and storage for packs and surgical carts. Nitrous oxide is sometimes included if nearby surgical rooms makes this cost-effective.

Debridement, contamination, and hydrotherapy rooms are usually equipped with non-slip waterproof floors, waterproof walls and ceilings, exterior type electrical outlets, open, high-volume, hand-controlled showers, ceiling-mounted, electrical patient rails and lifts, and physiological monitors (usually portable, battery-operated devices). HVAC requires special design attention for general humidity management, and debridement rooms need temperature management for the patients; ceiling-mounted heat lamps are often successful. Holding tanks for contaminated fluids may be required by local authorities.

Support Space

Adequate storage and easy access for linen, materiel, and equipment is essential. These spaces are to be designed to meet regulatory standards and must be easily accessible to the nurse in the flow of patient care activities.

Public Space

Public waiting space should be located near the critical care unit, accessible to elevators, rest rooms, drinking fountain, and phones, and not in the patient traffic flow. Separate access for visitors and patients into the unit is desirable. Grief and consultation rooms should be comfortably furnished and provide privacy.

FUTURE DEVELOPMENT

Because patient rooms and nurse stations could be in use for 20+ years, the planning team needs to address the possible future development of new devices and medical techniques and how to accommodate these future changes physically. General guidance includes the provision of adequate power and heat management. When a new or renovated facility is being designed, the team can also ensure that the provision of behind-the-wall backing plates for future mounts exist. Also, the team should ensure that spare, empty signal conduits, with access junction boxes and blank cover plates are provided in headwalls and side walls where charting or other activities may take place. The conduits can terminate in the overhead, near access hatches to allow for future addition of low-voltage signal cables.

EQUIPMENT PROCUREMENT PROCESS

From buckets to monitors, communication systems to x-ray viewers, all items must be present, stocked, and working on the opening day. There are many ways to obtain equipment. Some more accepted methods include direct, sole source negotiated purchase, group purchase agreements, competitive proposals, donations, and reuse of existing equipment. No matter how procured, the nursing staff expects the equipment to be appropriate, in place, and working. The method of purchase, while important, is highly dependent on several factors, including the types of items purchased, the system of purchase already in place, and is often dependent on informal relationships. Typically, devices needed for a project are grouped into purchase classifications dependent upon market source. If small in quantity and value, direct purchase through established distributors is typical. But, simple devices in aggregate for the project can amount to large dollar values, enough to be worth the efforts needed to assure the best value.

Request for Proposal

Procurement of complex, costly systems through the mechanism of a "request for proposal" (RFP) process is highly recommended. This is a flexible approach, allowing for negotiations and "fine-tuning" of final purchase terms. This can be especially important when the equipment to be purchased is highly visible, used by many staff members, and the technology is undergoing major changes. An example of these conditions is the planning for and procurement of physiological monitoring systems, but these considerations can apply to all complex equipment procurement. The flexibility of an RFP under the direct control and management of the hospital is of utmost value. The elements of a RFP consist of several major objectives: statement of purpose, hospital's terms and conditions, scope of work, technical specifications, quotation, and supporting documentation.

Statement of Purpose

The first section of a RFP should include a statement of overall purpose, or objectives, of the planning team. This is the section that can serve as a "fall back" position for recourse when equipment and systems "don't seem to be working quite right" after being delivered by the vendor. The following is typical of a lead statement:

"The vendor shall manufacture from all new materials, deliver, install, and demonstrate the proper and complete equipment and systems operations, from patient interface to information output, as specified in response to this Request for Proposal."

This section can also provide a general overview of the entire equipment system under consideration for instrumentation, the requirement that full staff training is to be included, that service support is expected for the "useful life of the system," and if the system does not function as promised in the proposer's response to the RFP, that the equipment will be removed at the direction of, and at no cost to, the hospital. A critical aspect for any serious proposer is the proposer's responsibility to be fully knowledgeable of the conditions of the site that will be present when the equipment and systems are installed.

Hospital's Terms and Conditions

Terms and conditions should be set by the hospital, not by the manufacturer. In one subsection of terms and conditions, each proposer must be notified that the needs of the hospital are paramount and that they drive the equipment specifications, thereby not allowing the manufacturer to reformulate the hospital's specifications to match his or her products. Further, the vendor's past performance at the hospital and other institutions will have a significant bearing in his or her success as a proposer. This subsection should contain a statement of the hospital's rights during and after the proposal process, the general guidelines to be used in selecting and negotiating with the successful proposer, and other aspects related to expectations of and protection for the hospital. Other subsections should obligate the proposer's response, product quality assurance, job execution, and other germane issues if the vendor is selected by the hospital.

Scope of Work

The scope of work is the hospital's opportunity to define the anticipated delivery, installation, and training schedule. An essential part is the definition of who actually installs what. This is a good section in which to provide a detailed description of each clinical area to be equipped by the successful proposer, providing details and number of beds, general description of conduiting, mounting, and other physical aspects that aid the proposer in scoping and pricing the RFP. If telemetry is included, this is where reception coverage can be defined. Since the vendor will deliver a totally functional system, the provision of an initial supply of disposables is desirable in addition to the other expectations defined in this section. Matters of hospital standardization, interfacing with other systems, and conformance to accepted standards could be appropriate here.

Critical to any system is the acceptance testing of the equipment and systems contracted. The vendor notifies the hospital when the system installation is complete and the final payment is due. When a RFP protects the hospital, staff has the right to verify that each element of the system hardware and performance meets specifications. Only the hospital should make this determination, and it should expect timely and complete corrections of all shortcomings. Control of final payment that is linked to acceptance of proper testing results and defect correction is often the key to an excellent system. Also, full documentation of testing results is a requirement before final payment.

Technical Specifications

This section is used to detail specific hardware and performance expected. Specifications can be divided into two sections: general and specific. The general technical subsection addresses visibility, legibility, alarm function, management of power failure, sequential loss of system function, software ownership and maintenance, and other system-specific concerns. The specific technical subsection addresses expectations of each parameter, display device, and printer/recorder functions, including range, precision, and accuracy for each.

Quotation

The information required and the format of the price response is critical in order to obtain an efficient, organized, and fair analysis of each proposer. These aspects are also important in eliciting alternative approaches and in establishing points of negotiation with the finalists.

This section can also be divided into several subsections: general quotation, systems quotation, unit costs, consumable cost, and service contract guarantees. The general quote subsection includes not only the overall price of the proposer's system delivered, installed, fully functional, accepted by the hospital, and training completed, but should include price guarantees and discounts if the entire system is obtained in a reasonable time period as well.

To allow flexibility and to allocate charges to various departments, a price breakdown by individual hospital department or patient care unit is typically required in the RFP. Multisystems discounts and the availability and cost of refurbished "demonstrator" equipment integrated into the systems should be asked for.

Unit costs further clarify aid in selection flexibility because the pricing of each device and major component is required. The cost of warranty is also priced out, as is installation, in the event that different budgets may be assigned these costs, or that possibly the work may be "brought inhouse." Warranty starts only upon acceptance of the final installation by the hospital.

Information is required for consumable items with guarantees of a price schedule through time, especially if the proposers identify that only their product can be used. This sometimes becomes an issue and should be addressed while the competitive bid atmosphere exists.

Service contract costs after warranty are also included. The future costs of spare parts, circuit boards, and components are requested. Critical to this section is the guarantee of the availability of parts for the useful life of the equipment and systems, and whether maintenance is under the vendor, inhouse, or through independent, third party providers.

Supporting Documentation

An appendix is usually added to RFPs to provide information about the project and to obtain details about the proposer. Reduced floor plans of the project are normally included. These can identify conduit size, location, and runs. Antenna coverage for telemetry systems are also delineated. This is also a place to list existing equipment that is ready for refurbishment and inclusion into the project.

Information requested from the proposer includes the organization type, officers, corporate and local addresses, and the name of the individual authorized to bind the proposer to the RFP. This mechanism is also used to list such factors as the proposer's teaching aids, location and size of service facilities, on-site loaner equipment, and spare parts. Sample maintenance contracts and warranties should be requested for review and negotiation.

Direct Purchase

Direct purchases occur when items are few in number, low in overall value, or purchase resources are limited. Purchases of this type are usually negotiated with established distributors doing business with the hospital at the time. It is in the interest of the distributor to provide reasonable prices in the expectation of continuing business with the facility in the future.

When aggregate device purchases reach significant values, it is usual to contact to several market sources in a more formal request for bid format. This provides an efficient means of assuring documented, low, and the most responsive cost rationale for the procurement of goods; it also provides proof of proper purchase techniques.

Group Purchase Agreement

Many institutions are members of group purchase organizations of various types. These entities are productive for virtually every level of purchase. Planning team attention should be directed to several important requisites when considering group purchase agreements. It is necessary to make sure that the exact device or system is obtained and that team specifications are understood and incorporated into the purchase. Also, the team must assure that the delivery is exactly what was expected. It is also essential that the purchase includes complete installation (when appropriate), training, manuals, and warranty. Although group purchase mechanisms are effective, when significant money is involved, the competitive proposal process is advisable.

Competitive Proposal Process

The proposal process achieves two major goals: it assures complete flexibility in developing complex system purchases, and it can produce actual cost reductions exceeding prearranged purchase agreements. The "discount" prices established by group purchase organizations actually become the floor or minimum discount for the equipment under consideration; experience shows that the proposal process produces significant discounts beyond prearranged agreement discounts. Two factors are critical for success in competitive processes. The process should be under the direct control of the hospital, and a competitive stance should be maintained when dealing with the manufacturing community.

Direct Control

This requires more involvement by hospital personnel than the simple group purchase, which is exactly why the planning team exists. The team develops procurement guidelines including elements critical to clinical, nursing, administrative, and maintenance personnel. Once developed, the guidelines must be effectively communicated to the manufacturing industry in a way that allows the hospital to maintain a highly competitive position to ensure not only that the best possible cost is achieved but that all needs and expectations are either met or exceeded.

Developing and maintaining a competitive stance is possible in virtually every situation. Guidelines are: (*a*) keep all internal discussion confidential; (*b*) speak to the outside world with "one voice" and never show internal dissension; (*c*) always state that dissension exists—but never be specific, only that it exists; (*d*) never exclude any manufacturer from consideration; (*e*) never show outward favoritism; (*f*) be very careful about off-site visits offered by manufacturers; consider using a basic site team equipped with complete agendas and written evaluation reports based on team guidelines, and if site visits to installations are made, be sure to include visits covering all proposed manufacturers; (*g*) conduct extended on-site demonstrations on patients with full staff training and without manufacturer presence after training. Elicit written staff evaluations for forwarding to the planning team.

Donations

Occasionally, donations of money or equipment occur. The former is most constructive, assuming that the donation is without restrictive conditions. Conditions are usually placed on the donation and each condition must be evaluated in terms of the lifetime cost of each restriction. The true value of a cash donation with restrictions may not be constructive for the project in the long run. Equipment donations can be more problematical; if accepted, the actual devices received may force operational compromises with lasting effects. Sometimes the planning team can work with the donator and turn the hardware offer into a cash donation.

One source of equipment donation is from manufacturers attempting to "enter the market," sometimes translating into gifts of up to virtually free systems. During preliminary evaluation phases, the planning team should evaluate these offers as if they were at "list price." If the system makes the final cut and the team has rejected all but the best two or three offers, then the "donations" become attractive and can be evaluated at the full gift offered, but with some cautions. If the manufacturer is entering the market, the system accepted may ultimately be the only one manufactured. A protection could be a manufacturer-paid performance bond replacing the entire system if certain market and service commitments are not met during the useful life of the system.

Other problems can occur with an established manufacturer entering the market with a new product line. Even though manufacturers carry out complete system design, product testing, clinical trials, and site testing, they can still offer new systems with problems, especially in the heat of competition, budget constraints, and the like. These factors should be considered when offered unusual discounts or significant "donations" and provide built-in protections.

Reuse of Existing Equipment

Reuse of existing equipment, possibly augmented by the purchase of similar, used equipment from other sources, can be a real enhancement to a new system purchase. The planning team should incorporate the reuse of existing equipment into all plans. Specifically, there are three approaches. First, the team should incorporate the refurbishment to "like-new condition" of existing equipment in the purchase of the new system. Second, the team should incorporate the replacement of existing equipment by new equipment, with the new equipment system purchase through a 1- or 2-year plan with guaranteed prices and service back-up. Third, before a final purchase agreement, the team should negotiate for trade-in of the existing equipment in exchange for credit or additional equipment with the purchase.

Timing

Projects take time. Large projects take several years. Procurement timing is always a factor under all circumstances. Because technological changes occur, and new medical services may develop within the hospital, a commitment to a manufacturer should not be made until the project is almost complete and opening is imminent. This requires current knowledge of manufacturers "lead time." Controlled procurement of equipment systems is critical for the success of the clinician's goals and objectives as well as the long-term success of a project. Suggested here is a three-step process:

1. Establish a well-formulated set of system performance specifications.
2. Maintain a very visible, competitive posture toward manufacturers.
3. Assure that the selected manufacturer installs an equipment system that conforms to the order given and that the system performs as promised.

SUMMARY

In the hospital setting, nurse managers are key members of any critical care building or remodeling project team and staff nurses are key participants in the planning process. Nurses translate family, patient care, and operational needs to the members of the planning team. To do this, nurses need familiarity with the process and an awareness of the vocabulary as reviewed in this discussion. The end result of physical plant design and equipment acquisition will thus be a "user friendly" environment.

GLOSSARY

Beneficial Occupancy. Formal turnover of the completed construction project to the "owner" by the contractor with the approval of the architect of record. This marks the date that staff can move in and fully utilize the new facilities. Ideally, the project is complete and perfect in every way. In practice, there is usually a "punch list" of incomplete items that the contractor will finish after the turnover. The punch list should have little impact on nursing function—an aspect for the planning team to watch.

Foot Candle. An accepted measure of visible light based on the illumination of a surface 1 foot away from a candle. There are more precise definitions but the planning team must remember that illumination is very subjective, subject to physiological accommodation, and therefore dependent on environmental conditions. The actual light level must be measured by appropriate meters.

Examples of illumination guidelines in foot candles (fc) on various tasks are: nursing corridor at night—3 fc; nursing corridor daytime —20 fc; writing surfaces—50 fc; work surface for normal tasks—50 fc; fracture table—200 fc; surgical field—2500 fc.

Junction Box. A box approved for electrical use, fixed to the building structure with attached conduits. Junction boxes may be used as "pull boxes" to aid in placing electrical wire or cables in long conduit runs. They are also used to make electrical junctions between wires and can contain equipment, terminals, connectors, and controls.

Punch List. A written list for items needing follow-up, changes, or corrections. There are several types of punch lists. Most punch lists will be created and maintained by the planning team. An example might be a construction item list containing corrections to be made by the architect and contractor. The architect may also provide the planning team with a list of questions or decisions needed for the project.

Other punch lists may include items to be addressed by hospital employees after beneficial occupancy, training schedules by manufacturers with levels of curriculum achieved by each staff member, and delivery of equipment for the project, etc.

The punch list should identify the originator, the entity responsible for action, the priority, the date first logged, the date needed, the date accomplished, and a review of satisfactory completion by the planning team.

It is essential for the planning team to separate clearly from all others those punch list items that are due under contract from the building contractor. Separation is necessary because these will go through the architect; it is important not to create extra work for the contractor, which may lead to change orders.

Rough-In. The physical space provided by the building contractor to receive and mount a device properly in such a way that allows for the device to function properly.

Considerations for rough-ins are dimensions, structures for attachment, proper locations and type of utilities (electrical power, steam, water, waste, ventilation), and finish on the rough-in surfaces. These considerations directly concern the hospital when a device is to be provided by the hospital for the contractor to install.

REFERENCES

1. Griebling ER, Pilcher S. Fire and safety codes in unit design. Dimens Crit Care Nurs 1984;3:98.
2. Hardesty T. Knowledge nurses need to participate on a design team. Nurs Manage 1988;19:49.
3. Baldwin A. Psychological transition to a new building. Stanford Nurse 1989;4:8.
4. American Association of Critical-Care Nurses (AACN). Standards for nursing care of the critically ill. Newport Beach, CA: AACN, 1989.

Chapter 11

Occupational Safety*

CLAIRE E. SOMMARGREN

The critical care unit is a marvel of modern technology. Equipment, machinery, and physical environment have been designed to meet every need of the critically ill patient. New diagnostic and therapeutic techniques are constantly being introduced. Unfortunately, this increase in technological complexity brings with it the increased potential for exposure to health hazards for both the critical care nurse and the patient.

It is recognized that there are actual and potential health hazards inherent in the critical care environment. Occupational safety issues in the health care setting have received increased attention in the literature, research, and regulatory arenas in the past few years. There is a renewed sense of awareness among critical care nurses that they must be vigilant to protect themselves and their patients from environmental hazards.

Workplace hazards are generally classified into five categories:

Biological;
Chemical;
Ergonomic;
Physical; and
Psychological (1).

Occupational diseases and accidents are potentially preventable. Nursing staff members and hospital administration must work together to implement effective environmental health and safety policies and strive for compliance with these policies in order to provide a workplace that is safe.

The first step in ensuring a safe workplace is the identification of hazards. A method frequently used by occupational safety personnel for surveying a workplace for potential hazards is the walk-through inspection. Persons conducting a physical survey actually walk through the unit and note as many hazards as possible.

Critical care nurses should become personally involved in hazard identification by being involved in formal walk-throughs and by conducting periodic informal walk-through inspections of their own. Because of their excellent observational skills, critical care nurses can become particularly adept at discovering health or safety problems and then can report hazards or potential hazards to the appropriate hospital department.

An essential factor in the prevention of occupational disease and accidents is an effective job training program. Orientation and continuing education programs for critical care nurses have traditionally addressed such areas as electrical safety, infection control, and body mechanics. However, to control workplace hazards adequately, the nurse manager and clinician must have a basic understanding of all known or suspected hazards and effective protective strategies. Critical care nurses must learn to view their environment, the devices it contains, and every aspect of their practice from a safety point of view.

Another way to prevent occupational safety and health problems is by substitution, that is, replacing the hazardous substance, procedure, or method with another that is less dangerous. An example of this is replacing a toxic cleaning agent with one that is less toxic.

Engineering controls also can reduce or eliminate hazards by modifying equipment or

---

*Some of the material in this chapter has been modified from Somargren CE, Carlisle PS, Heacock N, McCollam M, Rehm A, Travers P. AACN Handbook on Occupational Hazards for the Critical Care Nurse. Newport Beach, CA: AACN, 1988.

the workplace itself (2). Some examples are the installation of lead-lined walls in areas where exposure to ionizing radiation is expected, mechanisms that allow the quiet closing of doors, and ventilation systems that provide proper air exchange.

Some workplace hazards may actually be created by the work practices of some personnel. They may increase their own risk of exposure, or that of coworkers, by failing to adhere to safe work habits. For instance, a nurse may not dispose of needles properly, or may fail to report a frayed electrical cord before using a piece of equipment. Modification of these behaviors can be accomplished by education and frequent reinforcement.

Personal protective equipment, such as gowns, gloves, and goggles are the last line of defense in the control of occupational hazards. Adequate supplies of appropriate equipment must be available to staff members, and instruction in their proper use should be part of a comprehensive job-training program.

A strong employee health and safety program is an integral component in the protection of the health and well-being of the critical care nurse. Such a program should foster open communication between the critical care nursing staff members and those professionals with expertise in occupational health and infection control.

BIOLOGICAL HAZARDS

Infectious agents that potentially can be transmitted from patient to nurse have always been part of the critical care environment. These agents may be viruses, bacteria, fungi, or parasites. There has sometimes been a complacent attitude about biological hazards because the prevailing belief has been that anything could be taken care of by the appropriate antibiotic; however, recent concerns about human immunodeficiency virus and hepatitis B have renewed awareness for the need to control biological hazards.

Transmission or transfer of an infectious agent to a susceptible host can occur by several routes. Contact transmission, direct or indirect, can occur via droplet transmission, percutaneous/blood-borne transmission, skin and/or mucous membrane transmission, sexual contact, or oral/fecal transmission. The infectious agent may also be transferred via contact with a contaminated inanimate object. Other modes of transmission include: airborne transmission, that is, transmission by droplet nuclei or dust particles that can spread greater than 3–4 feet from the source; vehicle-borne transmission, such as contaminated water or food; and vector-borne transmission, such as insect spread.

Transmission of an infectious agent to a host may or may not produce infection or disease, depending on the susceptibility of that host. Contact with an infectious agent may cause only colonization of the host, that is, the presence of microorganisms that grow and multiply but do not gain entry to body tissues. Even though not infected themselves, colonized persons may serve as a source of infection to others. Infection occurs when the infectious agent gains entry to the body tissues and multiplies there. Infection may be symptomatic or asymptomatic. This infection may progress to disease, that is, the development of clinical signs and symptoms. Persons with asymptomatic infection may pose the greatest risk to critical care nurses because, while they exhibit no evidence of clinical disease, they are potentially infectious to those around them.

The nurse manager may have concerns regarding the assignment of pregnant nurses to infectious patients on the critical care unit. Studies have consistently shown that pregnant nurses are not at greater risk for infection than the nonpregnant nurse. It is of concern, however, that certain infections, such as rubella, hepatitis B, and cytomegalovirus, occurring during pregnancy have the potential for affecting the fetus and neonate as well as the mother. Because of this, it is very important that female nurses who are of childbearing age be well educated regarding potential infectious hazards and appropriate protective strategies, and that they incorporate this knowledge into their daily practice (3,4).

It is essential to take measures to prevent the transmission of infectious disease. There are a number of basic strategies which are effective in the control of all infectious agents. The first step in the control of these hazards is ongoing education. Staff members must have a good, up-to-date working knowledge of what biological hazards exist and how to control them. The hospital's infection control nurse should be considered an integral part of the critical care team and should be freely accessible as a resource to unit nurses.

Other important control principles that must be integrated into the nursing staff's practice at all levels include: removal of the biological hazard through washing of hands and skin; prevention of contact with the hazardous agent by covering the portal of exit or entry with such protective equipment as gloves, masks, eyewear or surgical dressings; and immunization. It is essential that the critical care unit be provided with appropriate handwashing facilities and adequate supplies of protective equipment. The staff must be encouraged to use such equipment whenever appropriate and not to feel the need to conserve these supplies at the expense of exposure to a biological hazard.

Hepatitis B Virus and Human Immunodeficiency Virus

Hepatitis B virus (HBV) and human immunodeficiency virus (HIV) are perhaps the most notorious of the biological hazards. They are discussed together here because their mode of transmission and control strategies are almost identical. Exposure to HBV and HIV occurs in the same way. The viruses are found in the blood, semen, and some other body fluids of infected persons, so that the primary mode of transmission is by infusion of contaminated blood, by percutaneous, mucuous membrane, or nonintact skin exposure to infected blood, and by sexual transmission. Of the two viruses, HBV has been found to be much more readily transmissable by all routes. This is presumed to be the case because the concentration of virus in the body fluids of persons infected with HBV is much higher than that of persons infected with HIV (5).

HBV infection can range from asymptomatic infection to mild, flu-like disease to fatal fulminant hepatitis. Clinical signs and symptoms include fatigue, anorexia, nausea and vomiting, jaundice, dark urine, clay-colored stool, hepatomegaly, and elevated liver enzyme levels.

HBV presents a serious threat to critical care nurses. Studies have shown that 10–30% of health care and dental workers show serological evidence of past or present HBV infection. The Centers for Disease Control of the U.S. Public Health Service (CDC) estimates that 12,000 health care workers become infected with HBV each year. Of these infected workers, 500–600 are hospitalized, 700–1200 become HBV carriers at risk of developing chronic liver disease, and 250 die, either from fulminant hepatitis, cirrhosis, or primary liver cancer (6).

It cannot be overemphasized that a significant number of patients are asymptomatically infected with HBV, and thus present a clear hazard to the nurse. This leads to discussion of a very effective control strategy—immunization. It is strongly recommended that all nurses practicing in high-risk areas, such as critical care units, be prophylactically immunized against HBV. A safe and very effective vaccine has been available for some time. The U.S. Department of Health and Human Services and the U.S. Department of Labor have stated that this vaccine should be provided at no cost to the worker (7). The nurse manager should urge critical care nursing staff members to become immunized. HIV infection may also remain unrecognized in patients in the critical care unit. Although persons infected with HIV may occasionally have a mild, flu-like syndrome several weeks after initial infection, they generally exhibit no symptoms at all for several months or years.

Infection with HIV results in selective defects in immune function. Although many elements of the immune system may be affected, HIV primarily depletes a subset of T-lymphocytes (8). The infection is first manifested by weight loss and persistent lymphadenopathy. Later, opportunistic infections and neoplasms may occur. HIV infection eventually leads to a defined illness referred to as acquired immune deficiency syndrome (AIDS) (9). AIDS is a progressive disease and recovery is not known to occur. Unfortunately, a vaccine offering protection against HIV is not yet available.

A control strategy that is essential in preventing exposure to both HBV and HIV is the implementation of "universal precautions" by which the blood and certain body fluids of all patients are considered potentially infectious for HIV, HBV, and other blood-borne pathogens. The CDC has provided clear guidelines on the implementation of these precautions in the clinical setting. These guidelines also contain recommendations regarding environmental control strategies, the management of HIV exposures, and the management of infected health care workers (10,11).

Strict adherence to universal precautions, the need to avoid needle stick injuries, and the care of hands so that the skin remains an intact barrier must be stressed. The nurse manager must follow the CDC guidelines, including the provision of adequate hand-washing facilities, appropriate waste disposal containers, and protective equipment supplies. The nursing staff must be encouraged to report immediately any percutaneous or mucous membrane exposure to blood or body fluids, and any such significant exposures must receive clinical and serological follow-up according to CDC guidelines and appropriate counseling.

Non-A, Non-B Hepatitis

Non-A, non-B hepatitis (NANB) is most often transmitted by transfusion of contaminated blood. Critical care nurses can be exposed to NANB hepatitis by the percutaneous route. The etiological agent has not yet been identified, and is presumed to be a virus (or viruses). Symptoms are similar to, but milder than, those of hepatitis B. Serological markers for other forms of viral hepatitis are absent. After the acute stage, chronic persistent hepatitis is common and persons are presumed to remain infectious. This form of hepatitis has also been reported to be transmitted via the oral/fecal route in water-borne outbreaks. Control strategies are identical to those for HBV and HIV. There is currently no vaccine available.

Herpes Viruses

Members of the herpes family of viruses are ubiquitous, and most persons have been exposed to one or more of them by adulthood. Those herpes viruses of concern to critical care nurses include herpes simplex virus I and II (HSV), varicella zoster virus (VZV), and cytomegalovirus (CMV). A distinct characteristic of these viruses is that they remain dormant in the body after initial infection, and may reactivate at some future time to cause active disease. Infection or reactivation of herpes viruses during pregnancy or the perinatal period may cause congenital defects or severe disease in the neonate (12).

HSV causes vesicular lesions of the mucous membranes and, sometimes, the skin. In labial herpes, these vesicles occur in the oral mucosa. Genital herpes is manifested by lesions of mucous membranes of the genitalia. A herpes condition which commonly occurs in nurses is herpetic whitlow, or paronychia, a painful infection of the area around the cuticle.

HSV is transmitted by direct contact with either the vesicular fluid in the herpetic lesions, or with oropharyngeal, respiratory, or genital secretions. HSV enters the new host via the mucous membranes or nonintact skin, as around torn cuticles. Control of transmission is very straightforward, and is focused on preventing contact with herpetic lesions or infectious secretions. Thorough frequent handwashing and the faithful use of gloves on both hands during suctioning, mouth care, taping of endotracheal tubes, and contact with lesions, cervical secretions, or vaginal mucosa can effectively prevent the transmission of the herpes simplex virus.

Varicella zoster virus (VZV) causes two distinct syndromes: chickenpox (varicella) and shingles (herpes zoster). Chickenpox is an extremely contagious but generally benign childhood disease characterized by fever, malaise, and a generalized vesicular rash. The major route of transmission is by respiratory droplets, but it is also thought to be spread by contact with vesicular fluid. Shingles is a reactivation of previous infection with VZV, and is manifested as a vesicular rash localized to one or more dermatomes. In immunocompromised patients, shingles occasionally will produce a generalized rash as in chickenpox.

Chickenpox is communicable via respiratory shedding of the virus from a time period approximately 48 hours before the rash appears. Thus, inadvertent exposure to chickenpox is not uncommon. In both chickenpox and shingles, communicability persists until the rash is dried and crusted (13).

Most persons have acquired immunity to VZV by the time they reach adulthood, either by having had chickenpox or asymptomatic infection. Approximately 90% of adults have immune antibodies when tested serologically. It is recommended that those nurses who have no history of chickenpox or shingles and are serologically negative for antibodies to VZV should avoid contact with patients who have VZV infection (14). Nurses who have been exposed and have no history of chickenpox or shingles should be tested serologically. If they are not immune, they are considered potentially infective during the incubation period (10–21 days after exposure) (15). These nurses must be restricted from direct patient care during that period. It is essential to report and deal with exposures to VZV very quickly to protect both staff and patients.

CMV is widespread in the environment. Initial exposure most often occurs during childhood, especially in day care centers, or during adolescence. Only occasionally do persons reach adulthood without previous CMV infection. Transmission of CMV usually occurs via droplet spread of respiratory secretions or direct contact with urine or respiratory secretions. It may also be transmitted through ingestion or sexual contact. CMV is commonly excreted by healthy, asymptomatic infants and toddlers, as well as immunocompromised patients. Transmission may occur by kissing, contact between mothers and their children, and contact between toddlers and their caregivers. Studies have shown that nurses do not acquire CMV at a higher rate than personnel having no contact with patients (16,17).

CMV infection is usually asymptomatic. Occasionally, it causes mild flu-like illness. During pregnancy, primary CMV infection or reactivation of previous infection can have harmful effects on the fetus. CMV infection is the most common congenital infection in the world, occuring in 1–2% of all live births; however, it causes symptoms in only 10% of those infants born with CMV.

Routine infection control practices are effective in preventing transmission of CMV for both the pregnant and nonpregnant nurse. Because it is not possible to identify those patients who are shedding CMV, these control strategies must be utilized with all patients. They include: routine hand-washing after contact with patients, body fluids, or respiratory secretions; use of gloves if contact with body fluids or mucous membranes is anticipated; refraining from kissing infants or toddlers.

Neisseria Meningitidis

Neisseria meningitidis (meningococcus) is a bacterium that is frequently present in the human oropharynx. It is generally harmless, and only rarely invades the body to produce disease. The bacterium is quite fragile and does not survive in the environment. It is transmitted by large droplet secretions, and is not airborne. Transmission from patient to nurse can occur only in situations where there is direct contact with oropharyngeal secretions, such as during mouth-to-mouth resuscitation or splattering during intubation or suctioning. Patients with meningococcal disease are considered to be noninfectious after 24 hours of appropriate antibiotic treatment.

Neisseria meningitidis can progress from asymptomatic colonization of the oropharynx to meningitis, meningococcemia, or pneumonia. Symptoms of meningococcal meningitis include headache, stiff neck, photophobia, and change in mental status. It is sometimes fatal. Symptoms of meningococcal pneumonia are similar to those seen in other bacterial pneumonias.

It is recommended that nurses wear masks when within 3 feet of patients with symptoms of meningitis until meningococcus has been ruled out as the causative agent. Transmission from patient to nurse is rare. However, it is essential to identify all health care personnel who may have been exposed and to assess the extent of exposure. In cases where that exposure has been significant, prophylactic rifampin must be provided within 48 hours.

INTESTINAL PATHOGENS

There are a number of organisms that can enter the human intestinal tract and cause disease. These are referred to as intestinal pathogens and commonly include *Salmonella, Shigella, Campylobacter,* and rotavirus. They can cause an unpleasant, but usually self-limiting, syndrome of nausea, vomiting, abdominal cramps, and diarrhea. In immunocompromised hosts, such illness can result in septicemia.

Transmission occurs by the oral/fecal route. Outbreaks are usually the result of food, water, or hands contaminated with infectious feces. Transmission from patients to nurses can be prevented by adherence to good handwashing procedures and the use of gloves whenever contact with feces or feces-contaminated objects is anticipated. It is also strongly recommended that nurses avoid eating, smoking, or applying cosmetics in patient care areas.

Hepatitis A Virus

Hepatitis A virus (HAV) is mentioned here because it is transmitted in the same way as intestinal pathogens. It causes hepatitis that is clinically indistinguishable from the other forms of hepatitis. Unlike other hepatitis viruses, however, the infectious agent clears completely after the acute stage. Asymptomatic carriage and sequelae do not occur. Diagnosis of HAV infection is made on the basis of serological testing. Control strategies are the same as for intestinal pathogens.

Mycobacterium Tuberculosis

Mycobacterium tuberculosis (MTB) is a bacterium that can cause infection at various sites in the body. Of these, only pulmonary and laryngeal infection are readily transmissible to the critical care nurse. This infection, commonly called TB, may be asymptomatic or symptomatic. Symptoms of active tuberculosis include fever, night sweats, weight loss, cough, and sometimes cavitation of the lung. However, it is the patient with unrecognized active tuberculosis (especially if intubated) who presents the greatest risk of transmission to the critical care nurse (18).

The bacterium is transmitted via the airborne route from the lung of a coughing or intubated patient, and inhaled into the lung of the nurse. When tuberculosis is suspected or confirmed, the patient must be placed in an isolation room with proper ventilation. If the patient is awake and cooperative, he should be instructed in the proper use and disposal of paper tissues. If this is not possible, a mask must be worn by personnel who enter the patient's room. By far, the most effective control strategy is the early recognition and treatment of infected persons. Because of an apparent association between TB and HIV, the CDC has recommended that HIV-infected patients be given a TB skin test, and begin appropriate medication if the test is positive (19).

Sarcoptes Scabiei

Sarcoptes scabiei (scabies) is a parasitic mite that is transmitted from skin to skin during direct contact with an infected person. The mite burrows under the skin, causing intense itching, papules, vesicles, pustules, and visible burrows. These skin disruptions appear on the forearms, hands, wrists, axillae, genitalia, areas beneath the breasts, and inner thighs.

Application of scabicide is effective in eliminating the mite, and may be repeated after 7–10 days. Prophylactic use of scabicide is not recommended, unless the nurse has had prolonged skin to skin contact with an infected patient. If it is not possible to treat an infected patient with scabicide immediately, gloves and gown must be worn whenever contact with the patient is anticipated. All personnel who have been exposed to scabies must be evaluated in order to prevent an outbreak on the unit.

CHEMICAL HAZARDS

There are subtle chemical hazards in the outwardly "safe" setting of the critical care unit. Substances that are used daily are potentially harmful to all health care workers. As with other workplace hazards, the key to control is knowledge. It is essential that all critical care nurses be knowledgeable about the nature and toxicity of chemical substances present in the critical care setting. Recent legislation underscores the need for employees to have adequate information about risks they may face. Many states have enacted right-to-know legislation regarding chemical hazards in the workplace. The Federal Hazard Communication Standard requires all employers, through training and education sessions, to inform employees about the substances with which they work, the health risks associated with those substances, and appropriate strategies to decrease or eliminate risk from exposure to these substances. Informational sheets, called Material Safety Data Sheets (MSDS), outlining specific chemical hazards present in that particular workplace, must be available to employees. The amount of information on these sheets is variable; some are more detailed than others. MSDS serve as a brief overview of chemical hazards, but often the information they contain is incomplete. They are a starting point from which nurses should be encouraged to find out more about the substances with which they work.

Exposure to chemical hazards may be acute, that is, of short duration, or they may be chronic, lasting several weeks to years. Whether or not adverse health effects occur following exposure depends on the concentration and form of the substance, the duration and frequency of exposure, work practices, individual susceptibility, general working conditions, and the presence of other exposures (20).

Exposure to a toxic substance can occur by three major routes. Inhalation is perhaps the most important route to consider, because it allows the most rapid rate of entry of a substance into the bloodstream. Skin absorption is a common route of entry that is made more rapid when the skin is chapped or abraded and no longer forms an intact barrier to toxic substances. Ingestion may seem an unlikely route of entry; however, it can result in significant exposure when nurses eat, smoke, or apply cosmetics after handling toxic substances.

Adverse health effects may be acute, that is, occurring soon after exposure. These effects may include headache, vertigo, skin rashes, or acute respiratory distress. Sometimes effects occur only after prolonged exposure. Symptoms may not appear until years after initial exposure to a toxic substance. Such chronic effects might be liver disease, chronic lung disease, cancer, or genetic defects.

Antineoplastic Agents

Antineoplastic, or cytotoxic, agents are administered in some critical care units. Antineoplastic agents are pharmacological substances used to disrupt or destroy cancerous cells in the human body. Nurses may be exposed to these agents during preparation, administration, or excretion of these agents. The routes by which these agents may enter the nurse's body are inhalation of drug dust or droplets, skin absorption, or ingestion through contact with contaminated hands, food, or cigarettes.

Antineoplastic agents have been reported to cause acute health effects such as vertigo, nausea, headache, and irritation or ulceration of the skin, mucous membranes, or eyes. Although research on these agents is incomplete, recent studies have suggested chronic effects involving carcinogenesis (causing cancer), teratogenesis (causing deleterious effects to fetal development), and mutagenesis (causing damage to genetic material) (21).

The Occupational Health and Safety Administration (OSHA) has published specific guidelines, "Work Practice Guidelines for Personnel Dealing with Cytotoxic (Antineoplastic) Drugs," regarding the safe preparation, use, handling, and disposal of antineoplastic drugs (22). This document outlines specific techniques and equipment that must be used with antineoplastic agents, such as biological safety cabinets (vertical flow containment hoods), appropriate types of gloves and syringes, and proper hygienic techniques. Only those staff members who have had special education and training should handle any aspect of cytotoxic drugs. The recommended equipment for safe preparation and handling must be made available in critical care units where antineoplastic agents are administered.

Other Pharmacological Agents

It is appropriate to mention the risk potential of exposure to a wide variety of pharmacological agents. Very little information is available on short- and long-term health effects of classes of pharmacological agents other than antineoplastics. It is advisable, however, for the critical care nurse to limit exposure to all of these chemicals as much as possible.

Compressed Gases

The compressed gas most frequently found in critical care units is oxygen, although helium and carbon dioxide may also be present in some units. The majority of oxygen used in modern hospitals is "piped in" from a central source. Cylinders of compressed gas, however, are commonly used during patient transport. The gas in these cylinders is compressed until it is almost liquid. If the cylinder is heated, the gas inside expands and causes increased pressure. If this pressure becomes too great, the cylinder can explode. For this reason, it is essential that compressed gas cylinders be stored away from sources of temperature extremes. These cylinders also present a safety hazard if they should fall and break open, causing a sudden escape of pressurized gas. Therefore, they should always be fastened securely to a wall or in a rack. If this is not possible, lay them flat on the floor until they can be removed from the area. During patient transport, cylinders should be carefully fastened to the stretcher or bed.

It is also important to remember that oxygen, while not a flammable gas, does accelerate the combustion of flammable material. Therefore, lighted cigarettes or open flames must be prohibited in areas where oxygen is in use.

Soaps and Cleaning Agents

A wide variety of soaps, detergents, and cleaning agents is found in critical care units. While these soaps and agents are necessary to maintain a clean and healthful workplace, they themselves may pose a health hazard to nurses. Most of the adverse effects of these agents are acute and short-term. Long-term effects are not yet well documented.

Common and immediate effects of soaps and cleaning agents are the drying, chapping, and abrading of the skin. This may be offset somewhat by the frequent use of a moisturizing lotion. Maintenance of skin integrity is extremely important in providing a barrier to the entry of both biologically and chemically hazardous material.

Occasionally, skin reactions are more serious. Some nurses may become sensitized to an agent over a period of time and develop an allergic reaction, manifested by inflammation, rash, or blistering. The MSDS may be useful in identifying the component in the soap or cleaning agent which is responsible for the allergic reaction. More than one kind of soap should be available so that nurses may avoid those that cause skin problems.

Some other agents frequently found in critical care units which can cause adverse effects are gluteraldehyde and household bleach. Gluteraldehyde and similar chemicals are sometimes used to clean medical instruments and equipment. They have been reported to cause skin, eye, and respiratory irritation, lightheadedness, and nausea. Areas where these chemicals are used must be provided with adequate ventilation. Bleach is used frequently in the disinfection of patient care areas. It is both a skin and respiratory irritant. The extent of this irritation depends on the concentration of the bleach. Prolonged skin contact and inhalation of mist from spray bottles must be avoided. Cleaning agents, like all chemicals, must be clearly labeled, including the name of the substance, its concentration, toxicity, and any safety hazards it may pose.

Solvents

Organic solvents are carbon compounds that are capable of dissolving other substances. Some solvents that are commonly found in critical care units are isopropyl alcohol, acetone, and benzoin. These substances can enter the body via skin absorption or inhalation. Only short-term symptoms have been documented thus far, and may include: skin irritation or drying, vertigo, drowsiness, headache, fatigue, insomnia, feelings of exhilaration, concentration or memory problems, and mood changes (23).

Control strategies focus on preventing skin contact with the solvent by wearing gloves, and on preventing inhalation by keeping solvent containers tightly covered when not in use. As with all chemicals, containers must be clearly and accurately labeled.

Waste Anesthetic Gases

The problem of waste anesthetic gases leaking from anesthesia equipment is well known in operating rooms, where scavenging equipment is used regularly to eliminate these gases from the air. This hazard has not, however, been frequently addressed in the critical care unit. Increasingly, critical care units are admitting patients immediately postoperatively, bypassing the usual large, openly ventilated, postanesthesia care unit. The source of waste anesthetic gases in the critical care setting is the air emitted from patients who have received inhalation anesthesia, whether they are breathing without assistance or on ventilators.

Studies have shown that even low-level exposure to waste anesthetic gases can cause toxic effects (24). Some reported adverse health effects include: nausea, fatigue, irritability, headache, impairment of perceptual, cognitive and motor skills, cancer, hematopoietic disease, liver damage, increased incidence of congenital anomalies in children of exposed women, and an increase in the risk of spontaneous abortion in exposed women and in the wives of exposed men.

Proper ventilation with adequate air exchange in patient care areas is the most important control strategy. Critical care nurses should avoid close or prolonged contact with the exhaled air from postanesthesia patients and should leave the area if acute toxic symptoms are noted. The nurse manager should pay particular attention to recurring reports of acute symptoms. Actions taken to prevent these short-term symptoms may avert more serious long-term effects in the future.

ERGONOMIC HAZARDS

The term ergonomics has its origins in the Greek words "ergon," meaning work, and "nomos," meaning knowledge (25). Ergonomics is a science that attempts to form a good match between the worker and the job. This can present quite a challenge in the critical care unit. Physically, units are usually crowded with a wide array of equipment which, inevitably, gets in the way of smooth working and traffic patterns. It is sometimes difficult for the nurse even to reach the patient through the maze of monitors, wires, and intravenous equipment. Working and lifting in awkward positions are more often the rule than the exception. Emergency situations may further compound the problems of this environment. Awkwardly placed electrical cords or inadvertent spills may result in falls. Added to this are the noise and stress that come with working in such a setting. Inappropriate lighting is often cited as a problem in critical care units. It tends to be either harsh and glaring or inadequately dim. Either of these may cause eye strain and increased stress levels.

Some of these problems, such as inappropriately sized patient care areas, can only be dealt with during the design or remodeling of a unit. However, there are specific ergonomic hazards that can be controlled more readily. The nurse manager must look critically at the unit to see if there are areas where environmental improvement can be made. Some minor health complaints among the nursing staff can be alleviated or eliminated by some simple changes in the work setting.

Lifting

Back pain is a major occupational health concern for critical care nurses. Almost one-half of all workers' compensation claims for hospital workers involve back injuries (26). Many factors may contribute to this problem, including the type of work being done and individual risk factors. Nurses most often associate episodes of back pain with patient lifting, which often must be done quickly and under less than ideal conditions.

Although it has been well studied, the etiology of back pain is not always clear. It may result from simple muscular strain, a ruptured muscle or tendon, or vertebral disk injury. It is generally thought that simultaneous lifting and twisting contributes to many back injuries. In addition, lifting creates high compressive forces on the vertebral disks, which may lead to their degeneration (27). Trauma may be a single, direct insult or the result of repeated injury (28). The commonly assumed association between poor body mechanics and back injury has not been supported by research.

Bedside critical care nurses are constantly called upon to lift and manipulate patients and equipment. Mechanical patient lifts and transfer devices are helpful in reducing stress to the back and should be provided and used whenever possible. Nurses must acknowledge physical limitations and should be encouraged to ask for sufficient assistance with lifting or moving patients. Nurses should be encouraged to participate in regular exercise programs in order to strengthen the muscles and other body structures involved in lifting. A regular back injury prevention/follow-up program that incorporates basic anatomy, lifting techniques, awareness of foreseeable hazardous situations, first aid techniques, home activities, stress management, and weight control has been recommended for both new and experienced staff (29).

Lighting

Lighting in the critical care unit must fit the type of work being done, the workers' comfort, and the time of day. Often, this is not the case, and the level of lighting is either too harsh and glaring or inadequate. Exposure to high intensity light and glare can result in short-term adverse effects such as ciliary muscle strain and iritis. This condition, called asthenopia, causes impaired vision with pain in eyes, the back of the head, and the neck (30). This is a reversible condition, and can readily be prevented by modification of the lighting level.

Lighting in the critical care unit should be designed to complement the type of work being done and to produce a comfortable and pleasant work setting. Provision should be made for lower level lighting during the nighttime hours, not only for comfort, but also to decrease the psychological effects of round-the-clock lighting for both patients and staff.

Work Station Hazards

The term "work station" refers to the immediate physical environment in which nursing care is provided. Hazards associated with this setting include anything that might cause contact with objects in the environment, interference with the normal pattern of work flow, or undue exertion or strain during the performance of routine tasks. Work station hazards may abound in the critical care unit. Frequently, the work space provided is of inadequate size or inconvenient configuration. Bulky or unbalanced equipment may tip over and fall. Furniture and fixtures, such as monitors, may be inconveniently located. Floors can be uneven, particularly in older units that have been remodeled.

Adverse health effect from work station hazards may include fatigue and injury to the soft tissues, particularly of the back, neck, and upper extremities. Symptoms such as these should be followed up to identify any contributing factors. Work station layout must be critically analyzed to eliminate design flaws which may be causing problems, and the nursing staff should be encouraged to report any unsafe conditions promptly.

PHYSICAL HAZARDS

Physical hazards can be defined as those environmental elements that may potentially produce harm through the transfer of physical or mechanical energy to other objects in the environment. Some physical hazards commonly encountered in the critical care unit are electricity, noise, and radiation.

Electrical Hazards

Most medical devices are electrically powered, and new equipment is introduced constantly. Electricity also presents a clear workplace hazard for both electrical injury and fire, and must be treated with great respect. It poses a potential danger not only to the critical care nurse, but also to the patient. While the focus of this section is on the nurse, the special susceptibility of some patients to electrical shock is briefly discussed.

Electrical current is the flow of electrons. Electrons flow only when a voltage difference exists between two points and there is a conductive path between these points. This voltage difference is generated at the power plant and is brought into buildings via power lines. Access to this voltage is through the three-hole wall outlet. The slot on the top right is the hot or charged line, the slot on the top left is the neutral line, and the round bottom hole is the ground line. Current wants to flow from the hot line to the neutral line, but it cannot do so until there is a conductive path. A piece of equipment plugged into the outlet provides such a path. The current passes through the equipment, powering it, and then back into the neutral line of the outlet.

When everything is operating properly, the current is contained, and there is no chance of exposure to an electrical hazard. Unfortunately, electrical equipment is never perfect, and some of the current can leak into other parts of the appliance, such as the outside case. This stray current is normally taken harmlessly to the ground line in the outlet. However, if there is a fault either in the equipment or the ground line, electrical charge may build up and present a danger for those using the equipment.

Metal and water are good conductors of electrical current. The human body also can conduct current, particularly internally through fluids such as blood. Therefore, the nurse or patient can inadvertently provide the conductive path between a faulty piece of equipment and a good ground, which may be another piece of equipment or a grounded metal bed. The result is a shock or electrical interference in monitoring equipment.

Sixty-cycle interference on cardiac monitors should be investigated promptly. Most often, the cause is poor contact at the skin electrode and is easily remedied. If this is not the case, the interference may be due to a faulty piece of equipment. Each piece of electrical equipment in the room should be unplugged one by one to identify which device is causing the problem. If a particular piece of equipment is identified, it should be immediately taken out of service and clearly marked as damaged. If the problem is not pinpointed, the appropriate hospital department, such as biomedical engineering, should be notified to investigate the cause of the interference.

Electrical current, that is, the rate of flow of electrons, is expressed in amperes. The health effects of electrical shock depend on the amount of electrical current that flows through the victim. These effects may be merely a tingling sensation, or burns (ranging from slight to severe), cardiac dysrhythmias, and death. The most serious injury occurs when the current is beyond the "let go range." This refers to the level of electrical current at which the muscles in the arms and hands contract, and thus prevent the victim from letting go. This spasmodic contraction may progress to paralysis of the respiratory and cardiac muscles, leading to exhaustion, respiratory failure, and fatal dysrhythmias (31).

One of the best defenses against electrical injury is a basic knowledge of the principles of electricity and the ability to recognize what constitutes a hazardous situation. It is essential to include these elements in staff education programs. One of the most frequent sources of electrical hazard is damaged equipment. Sometimes damage is obvious, such as cracked or worn electrical plugs or frayed wires. Plugs that become warm during use or equipment that makes unusual noises, smokes, or produces a burning odor is obviously in need of attention. Sometimes, however, damage is more subtle. Any electrical equipment that has been dropped or has had liquid spilled into it should be considered unusable until proven otherwise, even if it looks undamaged. All devices that are even remotely suspected of being damaged or functioning abnormally should be taken out of service immediately, clearly marked as unusable, and referred to the appropriate hospital department, such as biomedical engineering, for inspection.

Another potential cause of electric shock is improperly grounded equipment such as plugs with bent or broken grounding pins and multiple-outlet adaptors that are ungrounded. Equipment with two-pronged plugs, "cheaters" (ungrounded adaptors), and electrical devices owned by patients or personnel have no place in the critical care unit. New equipment must be checked for leakage of current before it is put into use.

Frequently, electrical hazards are caused by improper treatment of equipment. Overloaded wall receptacles can present both electrical and fire hazards. A common mistake is the removal of plugs from wall receptacles by pulling on the cord, rather than grasping the plug itself and pulling it straight out from the wall. This may cause separation of the cord from the plug, an obvious electrical hazard. The use of extension cords should be discouraged as much as possible. If extension cords are in a location where they can be tripped over, they present not only a safety hazard, but also an electrical hazard due to damage to the plug and wall receptacle. If it is absolutely necessary to use extension cords in an emergency, they must be taped securely to the floor with electrical tape.

A dry environment must be maintained around electrical devices. Electrical equipment must never be operated with wet hands. Puddles of water or other fluid should be removed immediately from floors, beds, and any other surfaces. This becomes particularly important when electrical cardioversion or defibrillation is being performed.

Critical care nurses should be aware that certain of their patients are very susceptible to accidental electrical shock. The patient who has an open conductive pathway to his heart, such as a pacemaker wire or pulmonary artery catheter, is considered to be electrically sensitive. An amount of electrical current that might be barely perceptible to the nurse may be enough to produce ventricular fibrillation in the electrically sensitive patient. This is referred to as microshock. A microshock of as little as 10 MA (milliamperes, 1/1000 of an ampere) can cause fatal dysrhythmia in the electrically sensitive patient, whereas it would take a flow of 100 MA to produce dysrhythmia in a person with an intact skin barrier (32). Some specific measures can be taken to protect these patients: rubber gloves should be worn when handling pacing wires, wires and any connections should be covered with a waterproof occlusive dressing or rubber glove when not in use, and the patient and mechanical devices should not be touched at the same time. Nursing staff members should investigate immediately any report from patients that they can feel shocks or tingling sensations from equipment.

Maintenance of an electrically safe critical care unit requires the joint efforts of the nursing and engineering departments. Nursing staff and administration must view all electrical devices as potential hazards and must actively pursue a program of electrical safety through education and surveillance.

The National Electrical Code (NEC) has been adopted by the Occupational Safety and Health Administration (OSHA) as a national standard in the prevention of electrical hazards (33). Along with state and local regulations, it can be the basis for a strong biomedical engineering maintenance program. Such a program must include a periodic check of all electrical equipment and wall receptacles for proper function and grounding. Whenever new equipment is introduced, it should not be used until it has been thoroughly checked for safety and current leakage, and the nursing staff members have been instructed in its use. The biomedical engineer should be considered an active member of the critical care team.

Noise

Noise is defined as loud, discordant, or disagreeable sound. It is usually measured in decibels (dB), which measure the intensity (amplitude), rather than the loudness, of sound (34). The decibel scale is logarithmic, that is, a sound that increases by 10 dB becomes ten times as intense and subjectively is perceived as being twice as loud. The United States Environmental Protection Agency considers sound levels greater than 35 dB at night and 40 dB during the day as "noise" in a hospital (35). Noise levels can vary significantly, but studies have generally identified noise as a common problem in critical care units.

Frequent or continuous high noise levels can cause hearing impairment, but this is not usually the case in critical care units. Noise in critical care units tends to be moderate background noise with short episodes of higher levels of noise. Background noise might include the sound of ventilators, suction apparatus, beeping monitors, bubbling chest drainage devices, and conversations. Generally, this noise is not higher than the 60-dB range. Episodic sounds such as ringing telephones, alarms on monitors or other devices, and bedrail adjustment may cause higher levels of noise, some reaching the 90-dB range. Added to this is the fact that noise levels in the critical care unit persist on a 24-hour basis with little variation.

Primary effects of noise in the critical care unit are physiological and psychological stress for both the nursing staff and patients. Attempts should be made to limit noise and create a more tranquil environment. This can be accomplished most effectively during the design or remodeling of a unit. Ideally, patient rooms should have walls and doors; there should be some sound barrier between patient care areas and service areas such as utility rooms and nurses' stations. In an existing unit, a noise control program should include a concerted effort to limit hospital staff conversations, to respond to monitor alarms promptly, to limit noise in handling supplies and equipment, and to keep doors shut whenever possible (36,37). Other control measures include the insulation of walls and equipment, installation of acoustical materials on walls and ceilings, and modification of door mechanisms to prevent slamming.

Radiation Hazards

Radiation hazards to critical care nurses include ionizing radiation and laser radiation.

Humans are exposed to ionizing radiation on a daily basis from many natural sources, such as radioactivity in rocks, radon in the air, and cosmic rays from outer space. This radiation does not present a health hazard when within certain limits. Nurses in critical care units are at risk for exposure to ionizing radiation from sources such as portable x-ray machines, fluoroscopy equipment, and body substances from patients who have had certain nuclear medicine procedures. This radiation can become a health risk in the workplace if not properly controlled. Employees in modern hospitals have a low life-time risk of adverse health effects due to radiation exposure (38).

Ionizing radiation can damage components of body cells, thereby altering or killing these cells. If alteration occurs within the genetic material of the cell, health effects in the form of clinical disease may not appear for years, as in radiation-induced cancer, leukemia, or genetic effects. Exposure to ionizing radiation has been linked to the development of lung and kidney fibrosis, cataracts, aplastic anemia, sterility, and radiation dermatitis. Prenatal exposure can lead to death of the fetus from leukemia, in addition to congenital anomalies of various organ systems, small head size, and mental retardation (39). The National Council on Radiation Protection and Measurement (NCRP) has issued recommendations for controlling ionizing radiation hazards, and these are enforced through state and federal laws. The NCRP sets maximum permissable dose (MPD) limits for both the general public and occupational personnel exposure to ionizing radiation. Exposure is measured in rems. For radiology personnel, the MPD is 5 rems/year. For nurses and other health care workers, the MPD is 0.5 rems/year. If there is actual or potential exposure to greater than 1.25 rems/year, a film badge must be worn (40).

Portable x-rays are performed frequently in the critical care unit. It is essential to give adequate warning to all personnel before any x-rays are done. More importantly, an attempt should be made to limit as much as possible the number of x-rays done in the critical care unit. If it is at all safe to transport the patient, it is preferable to perform these procedures in the radiology department, where engineering controls for ionizing radiation are already in place.

When a potential ionizing radiation hazard is not avoidable, it is possible to minimize exposure by following the principles of distance, shielding, and time:

Distance

The further away the nurse is from the source, the smaller the dose of radiation that will be received in a given time. It is recommended that nurses stay a minimum of 10 feet from the radiation source (41).

Shielding

No part of the body should be directly exposed to ionizing radiation. It is recommended that lead aprons covering both the front and back of the body be worn whenever the nurse must be closer than 10 feet to x-ray or fluoroscopic equipment in operation. For more continuous exposure, such as angioplasty, it is recommended that lead thyroid shields and lead glasses also be worn (42). Lead gloves should be worn if hands are to be in the direct field or area of high scattering. Adequate protective devices should be available to all personnel. Proper care and maintenance of lead aprons is often overlooked. They should always be hung on racks when not in use. If they are folded, the lead may crack and allow leakage of radiation. Aprons should be checked on a periodic basis, along with portable x-ray and fluoroscopic equipment, according to established standards.

Although patients who have received injections of radioactive material during diagnostic procedures present no external radiation hazard, their body wastes do contain radioactivity for 24 hours after the procedure. These patients must be clearly identified to all personnel. Care must be taken to avoid contact with body wastes by wearing disposable gloves when emptying and cleaning bedpans, urinals, and emesis basins.

Time

The longer the time of exposure to radiation, the greater the dose received. Exposure to radiation, even when wearing proper shielding devices, should be limited as much as possible.

Critical care nurses must receive instruction in the proper use and care of shielding devices. Only staff members who have received specific, appropriate training should operate radiology equipment such as fluoroscopes.

Lasers are used in the critical care setting for surgical procedures. The word *laser* is an acronym for light amplification by stimulated emission of radiation. Laser light is concentrated and intense, and it poses a potential risk for injury to the eyes and skin.

There are currently no OSHA standards for exposure to lasers. A federal regulation program is administered by the Bureau of Radiological Health of the Food and Drug Administration. The American National Standards Institute also has published detailed guidelines on laser safety in health care facilities (43). Lasers are classified based on their intensity, and therefore, the biological risk they pose:

Class 1: Minimal power; no biological damage; exempt from control measures.
Class 2: Low power; low level of biological risk; eye damage is possible if a person stares directly into the laser beam; must bear a precautionary label.
Class 3: Moderate power; moderate level of biological risk; damage to the eyes can occur with direct exposure to the laser beam.
Class 4: High power; high level of biological risk; damage to the eyes and skin is possible from either direct or diffuse exposure to the laser beam (44).

Laser units used in surgery are class 3 and 4 units. Adverse health effects from exposure to laser light depends on the class and wavelength (color) of the particular laser and on the degree of exposure. There may be damage to the retina, cornea, or lens of the eye. Thermal damage to the skin can range from reddening to blistering or charring. It should be noted that the electrical power supply for all lasers is high voltage, and care must be taken to avoid electrical shock.

A comprehensive safety and education program is necessary to prevent accidental injury. All personnel working with lasers must receive specialized training, which includes knowledge of the type of laser, potential safety hazards, and appropriate protective strategies. Laser equipment must be clearly labeled concerning its type and classification. Appropriate engineering controls to prevent exposure to the laser beam should be provided. The laser beam must be prevented from striking any reflective surface, and any flammable materials, such as paper, must be kept away from the beam. Personal protective equipment, such as goggles or glasses, must be available whenever laser equipment is to be used, and should be periodically tested and properly maintained. Hand protection must be used if the hands will be in the target field area.

It is advisable to institute a program of ophthalmic examinations for those involved in working with lasers. These exams are recommended before employment, if exposure is known or suspected or if the employee has eye complaints that may be related to exposure, and upon termination. Any accidental exposure or injury must be investigated and treated immediately.

PSYCHOLOGICAL HAZARDS

Psychological hazards refer to those emotional and environmental stressors that result in adverse physical, emotional, and behavioral consequences (45). The critical care setting abounds with potential psychological stressors such as high patient acuity, rapidly changing environment, complex decision-making, intense interpersonal communications, 24-hour staffing schedules, less than ideal working conditions, and inadequate staffing. It is easier for nurses to work successfully in the critical care environment if they are aware of the stressors that face them and are knowledgeable about control strategies that they can use in their daily practice.

Stress

Stress is the term used to refer to both the adjustive demands placed on the body and the internal biophysical responses of the whole body (46). Stress has been described as positive, "eustress," and negative, "distress." Because stress is an integral part of life that cannot be avoided, the goal becomes to balance the side effects of distress with elements of eustress. Whether a demand is viewed as negatively stressful depends on peoples' perceptions of the situation, their past experiences, the task at hand, their interpersonal relationships, and their individual personality characteristics (47). When stress is referred to as a psychological hazard, the negative aspects of stress are being addressed. Critical care nurses are at high risk for job-related stress due to the nature of their practice environment and patient population.

The health effects of stress are varied and may be manifested as physical, psychological, or behavioral symptoms. Because of this wide range of responses, these symptoms may not immediately be recognized as being a result of stress. Physical symptoms reflect the body's "flight or fight" response. They may include muscle tension, eating and sleeping disorders, tachycardia, headache, increased blood pressure and serum glucose, and increased susceptibility to illness. Persons under excessive stress are noted to exhibit inflexibility, negativity, decreased productivity, depression, avoidance behaviors including tardiness and increased absenteeism, decreased decision-making ability, and an increased accident rate. In addition, they may experience feelings of emotional exhaustion, helplessness, hopelessness, and powerlessness (48). Chronic high stress levels have been associated with physical conditions such as heart disease, chronic hypertension, increased cholesterol levels, digestive disorders, infections, and rheumatoid arthritis.

Adverse effects of stress can be controlled by a combined program of staff education, environmental manipulation, and counseling. Members of the nursing staff will be better equipped to deal with stress if they are able to recognize its symptoms in themselves and co-workers. It is also important to point out the role of relaxation techniques, physical activity, and good nutrition in controlling stress. It is desirable to provide an atmosphere that is supportive and that gives the nursing staff members control of their work environment. The nurse manager must look critically at policies within the institution that may be contributing to workplace stress. Finally, a strong employee assistance program, including skilled psychological counseling when necessary, should be in place for those staff members requiring intervention.

Shift Work

Numerous studies have demonstrated that shift workers, particularly those who do night or rotating shift work, experience adverse physiological, psychological, and sociological effects. These negative effects are thought to be the result of the persistent disruption of the daily circadian rhythm by the shift work (49).

The majority of adults are active during the day and sleep during the night, that is, they follow a regular diurnal pattern. Nurses who work the evening or night shift follow a nocturnal pattern, active during the night and sleeping during daytime hours. Those who rotate shifts do not follow a regular pattern. With either the nocturnal schedule or irregular schedule, the circadian rhythms are synchronized differently from those of persons who follow the diurnal pattern. It is generally found that shift workers sleep fewer hours than nonshift workers (50).

Physiological symptoms associated with shift work include digestive disorders, sleep disturbances, headache, fatigue, and nervousness. Psychological and sociological effects can include irritability, anxiety, and depression. Shift workers are reported to have higher incidences of family and sexual problems, higher divorce rates, and decreased activities involving social interaction.

It is possible to decrease the negative effects of shift work by designing work schedules that disrupt body rhythms as little as possible. Rotating shifts for longer blocks of time, for example, monthly rather than weekly, can be helpful. When moving a worker from one shift to another is unavoidable, it is preferable to rotate clockwise (day to evening to night) to decrease adverse health effects.

Some individuals experience more difficulty with shift work than others. It is recommended that workers with medical conditions requiring continuous therapy, such as diabetes, heart and circulatory diseases, and anxiety syndromes avoid alternating shift work (51). It should be recognized that some nurses cannot adjust their body rhythms to shift work and experience severe disturbances that can be confirmed by physical examination. Such health problems may warrant moving such employees to another shift (52).

Chemical Dependency

Chemical dependency refers to the inappropriate use of a substance to such an extent that a person's job, family, health, and social life are affected adversely (53). Substance abuse may take the form of alcoholism or inappropriate use of drugs. It is difficult to assess the frequency of chemical dependency among nurses because of the lack of consistent statistics.

Substance abuse is a complex problem involving not only prolonged intake of a chemical, but also a pattern of physiological, psychological, and sociological processes (54).

Symptoms of chemical dependency may include a wide variety of somatic complaints including colds, congestion, gastric disturbances, and vertigo. These symptoms may progress to absenteeism, inconsistent work patterns, memory loss, or blackouts. Economic or marital problems may result from the substance abuse.

The nurse manager must be familiar with the symptoms of chemical dependency in order to facilitate early recognition and intervention. Changes in job performance should be accurately documented. Frequently, there is denial of a dependency problem by both the impaired nurse and coworkers. Others may attempt to protect the abuser by making excuses or assisting with assignments. This behavior only prolongs the problem and possibly endangers the dependent nurse, coworkers, or patients. Suspected substance abuse must never be covered up. It is essential to refer the affected person promptly to the appropriate treatment service.

Violence

Violence refers to physical force used so as to injure or damage a person or thing. This definition can be expanded to include not only battery, that is, physical touching of a person either directly or with an object, but also verbal abuse and assault. Assault is a threat to harm another physically or an unsuccessful attempt to do so. Nurses may be exposed to assault or battery on the part of combative patients, patients' families or significant others, or intruders seeking drugs. The extent of the problem of violence against nurses is not well documented. It is thought to be significantly underreported for a number of reasons, including the paperwork involved and fears that the attack will be interpreted as performance failure on the nurse's part or as an act provoked by the nurse (55).

Adverse physiological health effects of violence include physical injury and a variety of symptoms such as headaches, appetite and sleep disturbance, body tension, and increased startle response. Psychological and behavioral effects include anger, fear, depression, self-blame, and altered interpersonal reactions (56).

Control strategies center on the provision of a safe and secure working environment with adequate provision for the control of violent patients and an efficient emergency assistance system. Attention should be paid to providing adequate security, particularly in stairwells and parking facilities.

Educational offerings can be helpful in teaching nurses the skills to recognize and defuse potentially violent situations (57). Nurses must be instructed to request assistance promptly if confronted with acute violence or physical danger. Any violent actions that occur in the workplace should be reported immediately. It is recommended that institutions offer a comprehensive employee's victim assistance program, which includes medical help, counseling, legal advice, and insurance and workers' compensation information.

CONCLUSION

An overview of some of the most common occupational health and safety hazards facing critical care nurses has been presented. Resources and supplemental readings listed at the end of this chapter provide more detailed information about particular hazards. Nurses must take the responsibility to become well informed about dealing with these hazards and the standards, regulations, and guidelines associated with them. This knowledge assists nurses in becoming active in formulating their institution's policies and procedures to ensure a safe and healthful workplace.

The author wishes to acknowledge Patricia Hyland Travers, MS, RN, COHN for her careful review and critique of the manuscript in its early stages of preparation.

REFERENCES

1. Sommargren CE, Carlisle PS, Heacock N, McCollam ME, Rehm AA, Travers PH (eds). AACN Handbook on Occupational Hazards for the Critical Care Nurse. American Association of Critical-Care Nurses, 1987.
2. National Institute for Occupational Safety and Health. Guidelines for protecting the safety and

health of health care workers. Publ. no. 88–119, 1988.
3. Valenti WM. Infection control and the pregnant health care worker. Am J Infect Control 1986;14(1):20–27.
4. Williams WW. Guidelines for infection control in hospital personnel. Infect Control 1983;4(4):326–349.
5. National Institute for Occupational Safety and Health. Guidelines for prevention of transmission of human immunodeficiency virus and hepatitis B virus to health-care and public-safety workers, 1989.
6. U.S. Department of Health and Human Services. Guidelines for prevention of transmission of human immunodeficiency virus and hepatitis B virus to health-care and public-safety workers, 1989.
7. Department of Labor/Department of Health and Human Services. Joint advisory notice: Protection against occupational exposure to hepatitis B virus (HBV) and human immunodeficiency virus (HIV), 1987.
8. Hamburg MA, Koenig S, Fauci AS. Immunology of aids and HIV infection. In: Mandell GL, Douglas RG, Bennett JE (eds). Principles and Practice of Infectious Disease, 3rd ed. New York: Churchill Livingstone, 1990.
9. Centers for Disease Control. Revision of the CDC surveillance case definition for acquired immunodeficiency syndrome. Morbid Mortal Weekly Rep 1987;36 (suppl no. 1s):1–15.
10. Centers for Disease Control. Recommendations for prevention of HIV transmission in health-care settings. Morbid Mortal Weekly Rep 1987;36 (suppl 2s):1–18.
11. Centers for Disease Control. Update: Universal precautions for prevention of transmission of human immunodeficiency virus, hepatitis B virus, and other bloodborne pathogens in health-care settings. Morbid Mortal Weekly Rep 1988;37:377–382,387–388.
12. Valenti WM. Infection control and the pregnant health care worker. Am J Infect Control 1986;14(1):20–27.
13. Whitley RJ. Varicella-zoster virus. In: Mandell GL, Douglas RG, Bennett JE (eds). Principles and Practice of Infectious Disease, 3rd ed. New York: Churchill Livingstone, 1990.
14. Patterson WB, Craven DE, Schwartz DA, Nardell EA, Kasmer J, Noble J. Occupational hazards to hospital personnel. Ann Intern Med 1985;102(5):658–680.
15. Patterson WB, Craven DE, Schwartz DA, Nardell EA, Kasmer J, Noble J. Occupational hazards to hospital personnel. Ann Intern Med 1985;102(5):658–680.
16. Adler SP. Molecular epidemiology of cytomegalovirus: viral transmission among children attending a day care center, their parents, and caretakers. J Pediatr 1988;112(3):366–372.
17. Lipscomb JA, Linnemann CC, Hurst PF, et al. Prevalence of cytomegalovirus antibody in nursing personnel. Infect Control 1984;5(11):513–518.
18. Williams WW. Guideline for infection control in hospital personnel. Infect Control 1983;4(4):326–349.
19. Centers for Disease Control. Diagnosis and management of mycobacterial infection and disease in persons with human T-lymphotropic virus type III/lymphadenopathy-associated virus infection. Morbid Mortal Weekly Rep 1986;35:448–452.
20. Travers PH. Application of toxicological concepts to the occupational history. AAOHN J 1986;34(11):524–529.
21. Rogers B. Health hazards to personnel handling antineoplastic agents. Occupational Medicine: State of the Art Reviews 1987;2(3):513–516.
22. Yodaiken RE. Work practice guidelines for personnel dealing with cytotoxic (antineoplastic) drugs. Washington, DC: Office of Occupational Medicine, Directorate of Technical Support, Occupational Safety and Health Administration, 1986.
23. Baker EL, Fine LJ. Solvent neurotoxicity: the current evidence. J Occup Med 1986;28(2):126–129.
24. Rogers B. Exposure to waste anesthetic gases. AAOHN J 1986;34(12):574–579.
25. American Association of Occupational Health Nurses. A comprehensive guideline for establishing an occupational health service. Atlanta, 1987 (section 3).
26. National Institute for Occupational Safety and Health. Guidelines for protecting the safety and health of health care workers. Publ. no. 88–119, 1988.
27. Levy BS, Wegman DH (eds). Occupational health: recognizing and preventing work-related disease. Boston: Little, Brown, 1983.
28. Biering-Sorensen F. A prospective study of low back pain in a general population: occurrence, recurrence and aetiology. Scand J Rehab Med 1983;15(2):71–79.
29. Riley E (ed). Protecting hospital employees through better ergonomics. Hospital Employee Health 1984;3(10):132.
30. Levy BS, Wegman DH (eds). Occupational health: recognizing and preventing work-related disease. Boston: Little, Brown, 1983.
31. Mylrea KC, O'Neal LB. Electricity and electrical safety in the hospital. Nursing 76 1976;6(1):52–59.
32. Cooper K. Electrical safety: The electrically sensitive ICU patient. Focus Crit Care 1983;9:17–19.
33. National Fire Protection Association. National fire codes. Vol. 6 and 7. Quincy, MA:1983.
34. Turner AG, King CH, Craddock JG. Measuring and reducing noise. Hospitals 1975;49(15):85–86,88–90.
35. U.S. Environmental Protection. Information on levels of environmental noise requisite to protect public health and welfare with an adequate margin of safety, Report no. 550–9–74–004, Washington, DC: 1974.
36. Hilton A. The hospital racket: how noisy is your unit? Am J Nurs 1987;87(1):59–61.
37. Turner AG, King CH, Craddock JG. Measuring

and reducing noise. Hospitals 1975;49(15):85–86, 88–90.
38. Patterson WB, Craven DE, Schwartz DA, Nardell EA, Kasmer J, Noble J. Occupational hazards to hospital personnel. Ann Intern Med 1985;102(5):658–680.
39. National Institute for Occupational Safety and Health. Guidelines for protecting the safety and health of the health care worker. Publ. no. 88–119, 1988.
40. U.S. Nuclear Regulatory Commission. United States nuclear regulatory rules and regulations (title 10, chapter 1, Code of federal regulations—energy. Part 20, Standards for protection against radiation). Washington, DC, 1984.
41. Hale J. A radiation safety primer for hospital nursing staff. Philadelphia: Hospital of the University of Pennsylvania, Department of Radiology, 1988.
42. Patterson WB, Craven DE, Schwartz DA, Nardell EA, Kasmer J, Noble J. Occupational hazards to hospital personnel. Ann Intern Med 1985;102(5):658–680.
43. American National Standards Institute. American national standard for the safe use of lasers in health care facilities. Toledo, OH: Laser Institute of America, 1988.
44. Stoner DL, Smathers JB, Duncan DD, Clapp DE, Hyman WA. Engineering a safe hospital environment. New York: John Wiley & Sons, 1982.
45. Sommargren CE, Carlisle PS, Heacock N, McCollam ME, Rehm AA, Travers PH (eds). AACN Handbook on Occupational Hazards for the Critical Care Nurse. American Association of Critical-Care Nurses, 1987.
46. Marks LN, Reed JC. Stress, part 1: Assessment of stressors in the workplace. AAOHN Update Series 1984;1(9).
47. Marks LN, Reed JC. Stress, part 1: Assessment of stressors in the workplace. AAOHN Update Series 1984;1(9).
48. Marks LN, Reed JC. Stress, part 2: The challenge of making stress a positive force. AAOHN Update Series 1984;1(10).
49. Jung F. Shiftwork: its effect on health performance and well-being. AAOHN J 1986;34(4):161–164.
50. Department of Health, Education and Welfare. Shift work and health (HEW pub. no. NIOSH 76–203). Cincinnati, 1976:57–58;149–150.
51. Bryden G, Holdstock TL. Effects of night duty on sleep patterns of nurses. Psychophysiology 1973;10(11):36–42.
52. Long RJ, Wilkinson WE. Employee assistance programs: extensions of the occupational health programs. AAOHN Update Series 1984;1(4).
53. Jefferson LV, Ensor BE. Help for the helper: confronting a chemically-impaired colleague. Am J Nurs 1982;82(4):574–577.
54. Engel F, Marsh S. Helping the employee victim of violence in hospitals. Hospital and Community Psychiatry 1986;37(2):159–162.
55. Lenehan GT, Turner ST. Treatment of staff victims of violence. In: Turner JT (ed). Violence in the Medical Care Setting: A Survival Guide. Rockville, MD: Aspen, 1984.
56. Moran JR. In: Turner JT (ed). Violence in the Medical Care Setting: A Survival Guide. Rockville, MD: Aspen, 1984.

SUGGESTED READINGS

American National Standards Institute, Inc. American national standard for the safe use of lasers in health care facilities. Toledo, OH: Laser Institute of America, 1988.

Centers for Disease Control. Recommendations for prevention of HIV transmission in health-care settings. Morbid Mortal Weekly Rep 1987;36 (suppl 2s).

Centers for Disease Control. Update: Universal precautions for prevention of transmission of human immunodeficiency virus, hepatitis B virus, and other bloodborne pathogens in health-care settings. Morbid Mortal Weekly Rep 1988;37(24):377–382,387–388.

Department of Labor/Department of Health and Human Services. Joint advisory notice: Protection against occupational exposure to hepatitis B virus (HBV) and human immunodeficiency virus (HIV), 1987.

Garner J, Simmons B. CDC Guideline for isolation precautions in hospitals. Infect Control 1983; 4(4):245–325.

Jacobson S, McGrath MH (eds). Nurses Under Stress. New York: John Wiley & Sons, 1985.

Kaab GM. Chemical dependency: Helping your staff. J Nurs Adm 1984;14(11):18–23.

Kelly M. The regulatory management of the chemically dependent nurse. National Council of State Boards of Nursing, 1987.

Levy B, Wegman D (eds). Occupational Health: Recognizing and Preventing Work-Related Disease. Boston: Little, Brown, 1983.

Lucey J, Baroni M. Herpetic whitlow. Am J Nurs 1984;84(1):60–61.

Mandell GL, Douglas RG, Bennett JE (eds). Principles and Practice of Infectious Disease, 3rd ed. New York: Churchill Livingstone, 1990.

Meth I. Electrical safety in the hospital. Am J Nurs 1980;80(7):1344–1348.

Mott PE, Mann FC, McLoughlin Q, Warwick DP. Shift work: The Social, Psychological, and Physical Consequences. Ann Arbor: University of Michigan Press, 1985.

National Institute for Occupational Safety and Health. Guidelines for protecting the safety and health of health care workers. Publ. no. 88–119, 1988.

Patterson WB, Craven DE, Schwartz DA, Nardell EA, Kasmer J, Noble J. Occupational hazards to hospital personnel. Ann Intern Med 1985;102(5):658–680.

Rom WN (ed). Environmental and occupational medicine. Boston: Little, Brown, 1983.

Sommargren CE, Carlisle PS, Heacock N, McCollam ME, Rehm A, Travers PH (eds). AACN Handbook on Occupational Hazards for the Critical Care Nurse. American Association of Critical-Care Nurses, 1987.

Stoner DL, Smathers JB, Duncan DD, Clapp DE, Hyman WA. Engineering a safe hospital environment. New York: John Wiley & Sons, 1982.

Sullivan E, Bissell L, Williams E. Chemical dependency in nursing. Menlo Park, CA: Addison/Wesley, 1988.

Travers PH. Toxicology: An overview of fundamental principles. AAOHN Update Series 1986;2(19).

Valenti WM. Infection control and the pregnant health care worker. Am J Infect Control 1986;14(1);20–27.

Williams WW. Guideline for infection control in hospital personnel. Infect Control 1983;4(4):326–349.

Yodaiken RE. Work practice guidelines for personnel dealing with cytotoxic (antineoplastic) drugs. Washington, DC: Office of Occupational Medicine, Directorate of Technical Support, Occupational Safety and Health Administration, 1986.

RESOURCES

Occupational Safety and Health Administration (OSHA)—23 states, Puerto Rico, and U.S. Virgin Islands have their own OSHA programs. The rest of the U.S. is regulated by Federal OSHA standards.

Regional offices: Boston, New York City, Dallas, Kansas City, Philadelphia, Atlanta, Chicago, Denver, San Francisco, Seattle.

National OSHA Information Office: (202)523–8148.
  OSHA ensures that employers comply with the provisions of the Occupational Health and Safety Act. It conducts workplace inspections and reviews records of previous inspections. OSHA provides information on current standards in response to requests.

National Institute for Occupational Safety and Health (NIOSH)
4676 Columbia Parkway, Cincinnati, OH 45226
(513) 533–8287
(also regional offices in Boston, Atlanta, and Denver)
  NIOSH conducts research and develops recommendations regarding occupational hazards for OSHA; it also investigates workplace hazards requested by workers or employers.

Centers for Disease Control, U.S. Department of Health and Human Services, Public Health Service
Atlanta, GA 30333
  The CDC collects statistics on hospital infection control programs. It also publishes guidelines for infection control in hospital workers.

American Association of Occupational Health Nurses (AAOHN) 3500 Piedmont Road, NE, Atlanta, GA 30305
  AAOHN is a professional and educational organization of registered nurses interested in occupational health issues.

Association of Hospital Employee Health Professionals (AHEHP)
P.O. Box 2029, Chula Vista, CA 92012-2029
  AHEHP is a professional and educational organization involved with health and safety issues in hospitals.

Association for Practitioners in Infection Control
505 East Hawley Street, Mundelein, IL 60060
(312)949-6052

Nurses' Environmental Health Watch, Inc. (NEHW)
33 Columbus Avenue, Somerville, MA 02143
  NEHW is a national nonprofit organization involved with the education of nurses and the public about environmental and occupational health concerns.

Part VI

HUMAN RESOURCES SPHERE

Chapter 12

Roles and Relationships

PAMELA F. CIPRIANO, GRANDEE R. HARDY

People are the most valuable resource in an organization. Their skills, insights, ideas, energy, and commitment can cause the rise or fall of an organization. In order to analyze an organization with all its subunits utilizing the human resource perspective, one must have an understanding of people, groups of people, and the complex relationships between the two.

MAJOR ASSUMPTIONS

To understand the human resources sphere of organizations' environments, major assumptions drawn from the theory and knowledge of organizational behavior need to be reviewed.

1. Organizations exist to serve the needs of their people. This assumption expressly states that humans do not exist to serve the organizational needs.
2. Organizations and people have a matching need for each other. Organizations need the ideas, energy, skills, and talent that people bring, thus creating the life of the organization. The people need salaries, work opportunities, careers, and the social exchange that organizations provide.
3. The "fit" between the person and the organization must be "right" or one or both will suffer. A work setting should offer the person and the organization mutually satisfying outcomes. Some organizations, however, have created environments that are alienating, dehumanizing, frustrating, and detrimental to their people. Talents may be wasted and lives may be distorted as a result. Under these circumstances, people may respond by devoting most of their efforts to beating the system or one another.
4. Both the person and the organization benefit when the fit is good. Under ideal conditions, an organization can be energizing, exciting, productive, and rewarding for its people. Its members are able to do meaningful and satisfying work while, at the same time, they can provide the necessary resources to the organization in order to accomplish its mission (1).

HUMAN NEEDS

Based on the assumptions of a human resources framework, the concept of need is important. People need elements from the environment in order to survive and develop. Nature and nurture theorists have an ongoing controversy concerning human physiological and psychological needs. The nature theorists advocate that human behavior is determined by biological and genetic factors. The nurture theorists argue that human behavior is almost totally determined by learning from one's past experiences.

There is, however, a consensus emerging around the thesis that behavior among and between people is the result of an interaction between heredity and the environment. A need in the nature/nurture interaction may be defined as a genetic predisposition to prefer some experiences over others. These preferences present needs that energize and guide behavior.

The concept of need provides a way to assess conditions that are both favorable and unfavorable for people in organizations. A simple theory of needs assumes that every person has needs and that there is some similarity among people and their basic biological needs. People behave in a specific manner in order to satisfy their needs. They become dissatisfied or frustrated when their needs are not met, but in an environment where there is a high probability of satisfying their important needs, people are likely to develop and grow.

Needs Theories

For the last 30–40 years, social scientists have studied human needs. The following section reviews four theories that relate to understanding relationships and can be applied in the critical care setting.

Maslow's Hierarchy of Needs

Abraham Maslow grouped human needs into five basic categories arranged in a hierarchy from "lower order" needs to "higher order" needs. The hierarchy is traditionally pictured in a pyramid as shown in Figure 12.1. Maslow contended that lower order needs dominate behavior when they are not satisfied and that higher needs become important only after lower needs are met. This theory applies to all human beings in any state in life.

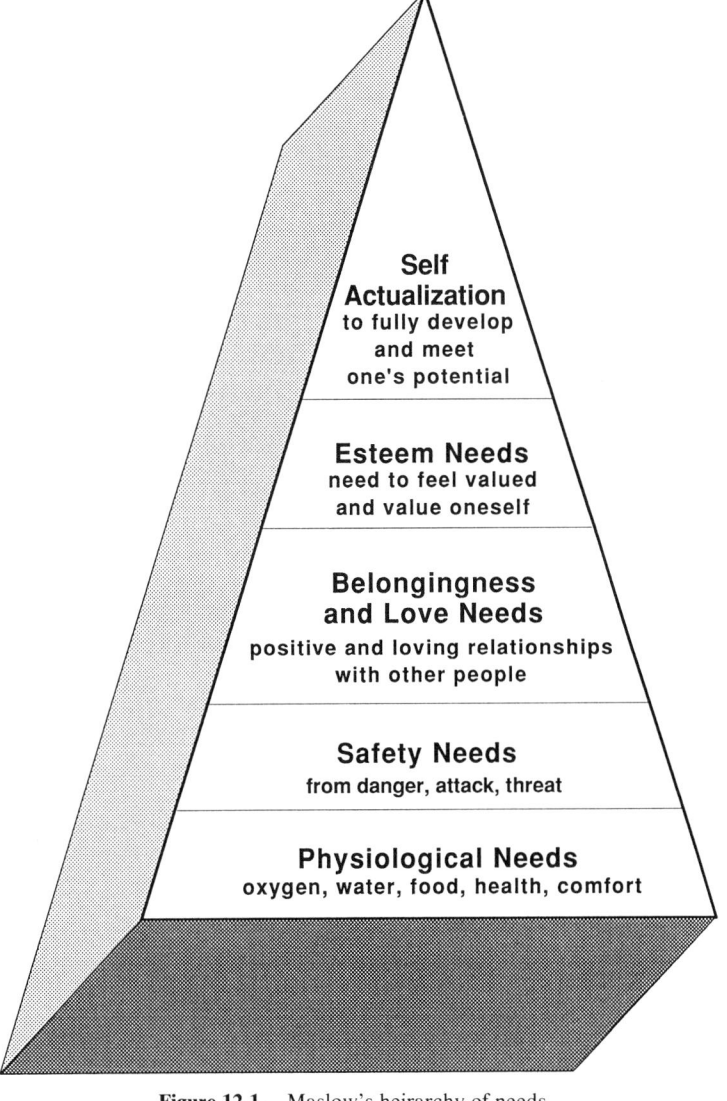

Self Actualization
to fully develop and meet one's potential

Esteem Needs
need to feel valued and value oneself

Belongingness and Love Needs
positive and loving relationships with other people

Safety Needs
from danger, attack, threat

Physiological Needs
oxygen, water, food, health, comfort

Figure 12.1. Maslow's heirarchy of needs.

McGregor's Theory X and Theory Y

McGregor's theory of motivation in organizational life is among the most widely read in management literature. The central theme of Theory X is that some managers actively direct and control the work of subordinates because they believe their subordinates prefer to be led, are self-centered, resistant to change, aimless relative to work ambition, and resist work. This theory has the manager treating subordinates as children. Theory Y purports that the essential task of management is to create an organizational environment and climate that is conducive to people's achievement of their organizational goals by channeling their efforts toward organizational rewards. The manager applying Theory Y relies on self-control and self-direction and treats people like adults.

Argyris Immaturity-Maturity Continuum

Another human resources theorist, Chris Argyris, outlined a basic conflict between human personality and organizational structure and management. He believed that people tend to develop in particular directions as they mature. People move from passivity to activity, from high levels of dependence to levels of independence, from having a narrow range of skills and interests to a more diverse range, and from a short-term perspective whereby interests are quickly developed and subsequently forgotten to a much longer term perspective in which the future can be envisioned and anticipated (2).

Argyris' theory also presents a conflict between people and organizations (3). Given the traditional principles of organizational design and management, employees are expected to be passive, dependent, and obedient at all costs. This may be observed in an organization with a rigid chain of command that requires people at upper levels of management to direct and control others at lower levels, thereby creating passive and dependent relationships. Task specialization with narrowly defined jobs creates the need for a chain of command to coordinate specialized work. Effective leaders, however, are sensitive to the process dynamics and enlist participation from the groups so that control may be loosened as employees demonstrate desirable behavioral outcomes and accountability.

Herzberg

Some of the best known work focusing on the individual employee is that of Frederick Herzberg, who studied work motivation and found that the causes of job satisfaction and dissatisfaction were not opposites but separate entities. He found that the satisfying factors in a job related to work tasks, whereas dissatisfaction related to the job environment. The conditions that are the dissatisfiers or "hygiene" factors include:

1. Salary;
2. Job security;
3. Working conditions;
4. Status;
5. Company procedures;
6. Quality of supervision;
7. Quality of interpersonal relations among peers, supervisors, and subordinates.

The intrinsic work conditions or job content are those components built into the job itself. This set of factors is called the satisfiers or "motivators" which include:

1. Achievement;
2. Recognition;
3. Responsibility;
4. Advancement;
5. The work itself;
6. The possibility of growth.

Considering human needs theories and a basic understanding of roles and relationships, the nature of management work in the critical care environment is based on the necessity to coordinate work and specialties. Critical care managers coordinate specialized work through application of planning, organizing, leading, and controlling others with a careful sense of balance. Management work focuses not only on the behaviors of organizations and their people, but also on the processes and infrastructures of organizations. The considerable interdependence of behaviors, relationships, and structure complicates the efforts of the critical care nurse manager.

CONCEPT OF ROLE

Individuals depend on others in their work group to help them learn and to set expectations of job behavior. Groups that work well together are able to understand each other's roles, communicate effectively, and experience changes in individual behavior to meet common goals.

Roles are related to task behavior. A role may include attitudes and values as well as specific kinds of behavior. It is what an individual must do to confirm the fact of his/her occupying a particular position. While a role is very broad, a position is usually very specific and differentiates among individuals. A position is a specific allocation of work activities and responsibilities (4). It is impersonal, has a specific title, and carries a set of formal and informal expectations. The activities and expectations of the position are written in a job description that is shared with the individual before employment commences. The same information describing the position should also appear in the evaluation criteria for the position and likewise should be shared with the individual. A job description also defines how one position relates to others in the same organization.

In critical care nursing units, it is common to have the majority of registered nurses holding similar positions, such as that of staff nurse. The staff nurse role is commonly understood as requiring major components of patient care, patient and family teaching, advocacy, communication, and delegation as well as supervision of ancillary personnel. The specific expectations of these role components may differ in job descriptions unit by unit and hospital to hospital.

Role Perception

Roles and positions are learned quickly and can result in major changes of behavior. It is important for the manager, the work group, and the organization to communicate clearly the expectations of a role and give feedback to one another.

There are different perceptions of the behavior associated with any role. In the workplace, there may be three different perceptions for any given role: that of the formal organization, that of a group, and that of the individual. The manner in which a role is perceived and the accuracy of this perception can have a definite impact on job performance

Organizational Perception

The position that an individual occupies in an organization is the sum total of roles as outlined by the organization. This includes defining the position in the chain of command, the degree of authority associated with the role, and the duties and expectations of the position. These role components are established by the organization and relate to the position and not to any particular individual or personality. For example, the critical care nurse manager role in some institutions may be a first line management position that reports to the critical care clinical director. This nurse manager has authority to manage the critical care staff nurses as well as clerical and technical staff within the unit. The scope of responsibilities includes the coordination of available resources to provide critical care nursing efficiently, effectively, and of a quality consistent with established standards.

Group Perception

Relationships develop between roles that link individuals to the various formal and informal groups to which they belong. Therefore, group expectations evolve over time and may not be congruent with the organization's perception of the role.

Individuals rarely see themselves relating to "the organization." Most of the time, they relate to one another and often in the context of groups. Work groups, committees, task forces, and project teams are often the vehicle for accomplishing a variety of tasks in critical care nursing. Managers, in particular, spend virtually all of their time in interpersonal exchanges. The role of the manager may also vary depending on the group activity. For example, the role and behavior of the critical care nurse manager may be quite different as a critical care committee member versus serving as the chairperson of the unit-based quality assurance committee.

Individual Perception

Every individual who holds a position within an organization or group has a clearly defined perception within one's mind. This perception is influenced by background, education, and social class. These influences coalesce to form the individual's basic values and attitudes that are brought to the workplace and, in turn, affect the individual's perception of his or her role. For instance, the leadership style and the manner in which a critical care nurse manager fulfills that role may depend on that particular manager's family background and culture, birth order, basic values of authority, and responsibility, as well as education.

Role Set

The critical care nurse occupies many roles at one time within the work setting. The expectations of behavior for a position at work are derived from a variety of sources. The different sources make up what is called the role set. The more and varied the expectations of an individual, the more complex the role set (5). For the critical care staff nurse, the role set is quite complex and is primarily composed of the patient and family, self, peers, supervisor, physicians, and other caregivers with whom the nurse regularly interacts. For the critical care nurse manager, the set is enlarged to include more administrative personnel, individuals in other departments, and staff in the unit. The expectations of each of these groups may be very different and are affected by such variables as perceptions, background, experience, values, and attitudes.

Role Conflict

Role conflict occurs when an individual is faced with simultaneous occurrence of two or more role requirements for which performance of one precludes the performance of others. There are basically three types of role conflict: person-role, intrarole, and interrole conflict (6). Table 12.1 provides examples of these types of role conflict within the critical care environment. Research has shown that role conflict leads to consequences such as increased psychological stress and other emotional reactions. Whenever possible, managers should seek to minimize role conflict and should be routinely looking for the presence of such conflicts that may be responsible for ineffective performance by individuals and groups.

Work Group

In critical care units, in addition to the care each nurse delivers to assigned patients, work is accomplished in groups or teams. No matter how large or small the critical care unit staff, group presence has an effect on each member's behavior. When members of a group respect one another, like each other, and have cohesiveness, they are usually more productive, have more influence over each other, and communicate more effectively than group members who do not share that cohesiveness.

Informal work groups can dramatically influence the behavior of group members and can directly affect productivity. Sociologist George C. Homans explored group work in order to explain how groups acquire the power to control behavior (7). He defined the three elements of a social system as activities, interaction, and sentiments. Activities are the tasks performed by individuals, interactions are the behaviors that occur between individuals performing tasks, and sentiments are the

Table 12.1. Types of Role Conflict

Type	Definition	Example
Person-role	Occurs when role requirements violate the basic values, attitudes, and needs of the individual occupying the position	Critical care nurse manager terminates a staff nurse who has a family for whom she is the sole earner
Intrarole	Occurs when different individuals define a role according to differents sets of expectations making it difficult if not impossible for the person occupying the role to satisfy all of them	Critical care nurse manager as staff advocate as well as a link to top management—the precarious "middle" position
Interrole	Occurs when an individual simultaneously performs several roles, some of which have conflicting and sometimes contradictory expectations	Physician as medical director of an ICU—role may call for decisions related to denial of admissions or cancellation of elective surgery due to bed unavailability

attitudes that form between individuals and within groups. He proposed that these concepts, although separate, are mutually dependent upon one another. Any change in one element effects some change in the other. The relationship between the elements is shown in Figure 12.2.

It is normal to interact when performing work. Subsequently, people develop sentiments related to satisfaction with their jobs and about each other. Increased interaction usually yields greater positive sentiments which, in turn, increases interaction. A cycle develops until a point of equilibrium is attained. In the process, staff nurses usually begin to act more alike and share similar sentiments. A result of this process is the development of group norms. The more cohesive the group, the greater the desire of individuals to conform. This powerful force can be directed toward accomplishing organizational goals. As a group develops, it is necessary to clarify roles and relationships. The opportunities for input, feedback, communication, delegation, control, and influence within each role in the group must be known. This information is generally available in job descriptions with additional clarification provided by the nurse manager.

In addition to nursing personnel in the work group, individuals in other roles interact frequently in the course of delivering patient care. These include individuals in assistive care-giving, technical, clerical, medical, and administrative roles. All the individuals in these roles share a common goal. It is necessary to recognize and accept the uniqueness and contributions of each different role in order to strive for positive outcomes.

ORGANIZATIONAL RELATIONSHIPS

The inevitability of groups' being the vehicle for accomplishing most work and a multiplicity of tasks has spawned a tradition of research on group dynamics. A number of theorists have emphasized that groups operate at two different levels: (*a*) an overt approach focused toward task accomplishment; and (*b*) a more subtle level of group maintenance and interpersonal dynamics. Group effectiveness depends mostly on who the members are, what resources they bring, and how well they are able to work together (8-10).

As groups work together in a cooperative manner, the result is collaboration. It is through collaborative efforts that multidisciplinary needs of patients are met. The critical care work group consists of all personnel regularly assigned to contribute to the care of patients and families within the unit. This includes nurses, other direct caregivers, and those who provide indirect care through technical and clerical support of unit activities. In addition, the work group takes into account individuals from departments other than nursing who routinely interact with the patients and staff members in unit. For the most part, the registered nurse staff members are considered the pivotal members of the work group since they perform the majority of care activities and maintain accountability.

The work group establishes relationships within the unit and outside the unit. Relationships within the unit are described as intraunit and involve only members of the work group itself. Interunit relationships are those that form among units within the nursing organization. External relationships are normally de-

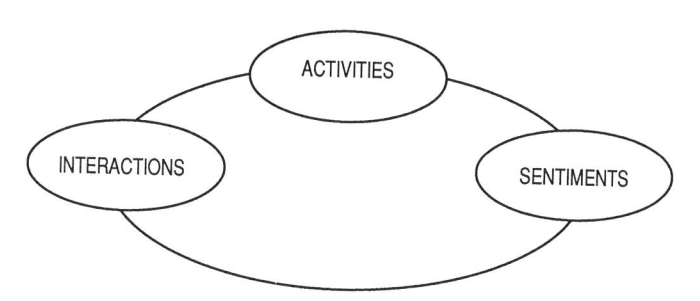

Figure 12.2. Relationships between elements.

fined by organizational structure and arise out of formal as well as informal associations.

Internal Relationships

Internal relationships within nursing can be within the unit or the organization as a whole. It is important to note that teamwork occurs not only at the unit level but also across nursing units within an organization. This collaboration may not always be readily apparent, however, and the critical care nurse manager communicates these collaborative endeavors to the staff members. The need for careful resource allocation influences integration and collaboration among critical care units and other inpatient units within hospitals.

Intraunit Relationships

Within the critical care unit there are informal alliances or groups among staff. These groups occur naturally along lines of similar attributes. They normally occur around such similarities as experience level, status or title, special skills, tenure within the unit, loyalty to new versus old management, shift rotation, and those who consistently express pro versus con opinions on unit issues. In a large group (greater than 10 individuals), it is expected that not all members necessarily like each other socially. It is important, however, that a healthy respect be cultivated among all members and that the basic expectation is that all members of the group work together to accomplish the common goals of the unit.

As previously mentioned, the uniqueness of each individual, strengths, talents, and contributions affect the way each person fulfills the role he or she performs. Finding ways to integrate all staff members can build cohesiveness by focusing on the positive qualities each member of the work group has to offer. Inviting diverse viewpoints can enhance problem resolution and effect decisions that more closely represent all members of the group.

Interunit Relationships

Interunit relationships are also an important part of a critical care unit's functioning within the nursing organization. If the hospital has only one critical care unit, the interunit relationships form with acute care and specialty units. Admission and transfer of patients are the common opportunities for dialogue and understanding of each unit's specific responsibilities in the care of a patient. When more than one critical care unit exists, there are often more complex issues that center around the differences in the units: the size, staffing, acuity of patients, expertise of staff, resource sharing, competition, image, and other potential areas of conflict. The key to productive relationships is building on positive attributes and recognizing that each unit provides a service and strives for the common goal of patient care and organizational effectiveness. The nurse manager is the primary contact with these other units and represents the critical care unit and its concerns.

External Relationships

The nature of hospital work requires that all personnel maintain formal or informal relationships throughout the organization. Cooperative efforts must occur to facilitate care in a hospital.

Interdepartmental Relationships

The functioning of the critical care unit is interdependent with the functioning of such departments as the laboratory, social services, central supply, pharmacy, respiratory therapy, laundry, and dietary. It is important for nurses to cultivate working relationships with personnel in other departments for the purpose of meeting patient care goals. Personnel in departments without direct patient contact may be unaware of the urgency and magnitude of the needs that arise in critical care. Nurses are charged with considering the "bigger picture," using creative solutions when possible, and articulating immediate needs by stressing urgency without exaggeration when seeking cooperation in a difficult situation. When an emergent need is not met because of lack of understanding on the part of personnel in other departments, a history of a working relationship increases the probability that productive problem-solving can be entered into immediately.

Physician Relationships

In critical care, relationships with physicians are crucial in the delivery of patient care and coordination of unit activities. The strong nurse-physician interdependence demands close communication and frequent interaction to clarify responsibilities and coordinate patient care. The structure of designated physician coverage varies by hospital organizational structure. Some critical care units have full-time directors who have total control of patient admission and oversight while others may be designated directors who share the responsibilities for medical direction among physicians within the institution. Whatever the arrangement, the quality of patient care outcomes depends on positive interaction and coordination of the critical care team. Knaus and colleagues (11), in a 1986 study, evaluated outcomes from intensive medical and nursing care in major medical centers and found that the "highest quality of care (. . .) appears to require a high degree of involvement by both dedicated physicians and nurses in ongoing clinical care." In facilities with better than expected mortality rates, the study also cited excellent communication between physicians and nurses as integral to meet patient care needs. The other factors that described successful nursing attributes at the facilities with more desirable patient outcomes were an increased independence in roles of staff, a comprehensive nursing education support system, and the joint establishment of patient goals by nurses and physicians.

Within the critical care unit, the designated physician director or responsible supervising physician works with nursing staff members and nursing leadership to make decisions regarding bed control and patient admissions, triages patients in emergency situations including staffing inadequacies, and oversees the quality of care delivered by all physicians and medical students interacting with patients in the unit. The physician director also acts as liaison from the physician staff and interprets medical decisions and actions to the unit staff members and leadership.

MODELS OF COLLABORATION

The nurse collaborates with individuals and departments to meet patient needs. When interacting with individuals who provide direct patient care to patients and families, there is a sharing of information and decisions among all parties as diagramed in Figure 12.3. Collaboration with other professionals commonly occurs when it is related to planning of patient care. In addition to patient benefits, a collaborative practice system with physicians has been associated with increased autonomy and decision-making, clearer delineation of professional and technical nursing roles, and improvement of job satisfaction for registered nurses (12).

When the nurse collaborates with individuals and departments that facilitate the delivery of prescribed care, the nurse acts on the patient's behalf in communication and interaction with others. This is represented by a *solid line* between the nurse and patient and a *dotted line* between the patient and others as seen in Figure 12.4. The nurse depicted in these models may be the staff nurse, the nursing case manager, the clinical nurse specialist, the nurse manager, or a higher level nurse administrator who acts as an advocate for the patient or family.

MODELS OF COLLABORATIVE RELATIONSHIPS

Synergy must occur in the workplace in order to produce predetermined goals and outcomes. The idea is to make the whole greater than the sum of its parts, or the productivity of the organization as a whole, greater than the total efforts and productivity of all individuals. Synergy occurs when people work together as a team. Teamwork is not only important for organizational productivity but is also necessary for individual fulfillment. Being a part of a

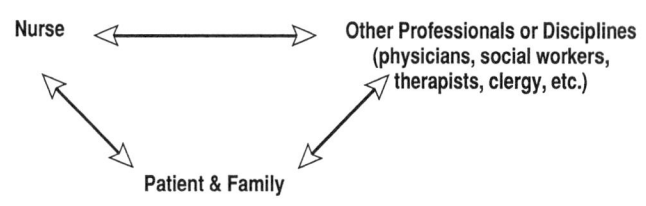

Figure 12.3. Sharing of information and decisions.

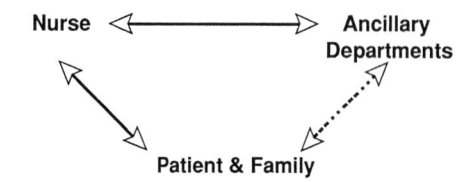

Figure 12.4. Nurse acting on the patient's behalf.

team, counting on other people, and being able to share ideas, problems, and challenges are inherent human needs.

Teamwork exemplifies the process of collaboration. This process entails problem-solving and reciprocal enhancement of the other's power as well as one's own. Thus, collaboration means team members sharing as much information with each other as possible and communicating through cooperation.

Collaborative Roles in Critical Care

Collaboration in health care has received an increased amount of attention over the last decade from many governing bodies of professional organizations as well as accrediting and regulatory agencies. The American Nurses Association and the American Medical Association worked together through the National Joint Practice Commission (NJPC) to produce "Guidelines for Establishing Joint or Collaborative Practice in Hospitals." The NJPC recognized that nurse-physician relationships are critical to producing desirable patient outcomes and wanted to demonstrate successfully how to improve these relationships so that patients, nurses, physicians, and hospitals benefitted from the results (13). The National Commission on Nursing, which studied nursing-related problems and issues in the United States in the early 1980s, encouraged collaboration by recommending in its action plan that nurses, physicians, health care administrators, and hospital trustees "promote and support effective nurse-physician interaction in clinical settings through policies that engender cooperation in patient care and a climate that fosters mutual respect and trust" (14). The Joint Commission on Accreditation of Healthcare Organizations (JCAHO) emphasizes collaboration by mandating certain multidisciplinary committees and work groups such as critical care quality assurance activities (15). The American Association of Critical-Care Nurses (AACN) and the Society of Critical Care Medicine collaborated in the development of organizational principles that support and enhance cooperation in the critical care environment. These principles have been published as AACN's Position Statement on the "Collaborative Practice Model: The Organization of Human Resources in Critical Care Units" (16). The document is reprinted for reference in Appendix B.

While there may be many ways to depict and describe working relationships, the model described by Conner and Palmgren (17) is quite useful. They describe three basic types of working relationships.

$$1 + 1 < 2 \quad \text{Self-Destructive}$$
$$1 + 1 = 2 \quad \text{Static}$$
$$1 + 1 > 2 \quad \text{Synergistic}$$

The self-destructive or negative competition relationship ($1+1<2$) describes a situation in which two people or subunits within an organization interact so that they consume resources at a greater level than they create outcomes. The result is less productive than if they had worked independently. An example within a critical care unit might be two senior staff nurses working in opposition to one another, both protecting their status and turf, refusing to communicate, and blaming each other for the deficits in the working relationship. They resemble two competitors rather than two professional colleagues. They actually have a negative production level.

The static relationship ($1+1=2$) is one in which both parties consume resources at the same rate of contribution to the system. The work output is predictable and is similar to working independently rather than combining efforts for greater productivity. This relationship works well initially; if, however, unanticipated stressors, conflicts, or events upset the balance to the slightest degree, the relationship can revert to negative competition. Static relationships have no reserves from which to draw to meet unpredicted needs or demands.

In a synergistic working relationship ($1+1>2$), people within the system interact so that their output exceeds their resource consumption. By working together, the result is productivity greater than if the two team members had worked independently. A critical care nurse manager who encourages and develops the skills of each staff member toward this end finds that the participation is offered freely and that it contributes to gains in unit efficiency and effectiveness such as cost-containment, nursing retention, clinical innovations, research, and open communication.

Success in the collaborative process requires that stereotypes be ignored and all ideas be given consideration based on merit, regardless of their source. This concept is important in the critical care setting where many different and specialized professionals and technicians as well as many support and ancillary personnel come together to care for the critically ill patient. Nurses in a variety of roles, assistive personnel such as patient care assistants, nursing assistants, monitor technicians, and unit secretaries, as well as the physician, all have both a direct and an indirect impact on patient care. Other departments such as respiratory therapy, laboratory, dietary, and radiology also offer services that are an integral part of critical care. It is imperative to the establishment and maintenance of synergistic relationships to recognize and accept the individual and unique contributions of each role in order to achieve the desired outcomes.

Critical Care Staff Nurse

The critical care staff nurse is a licensed professional responsible for ensuring that the critically ill patient receives optimal care by demonstrating professional accountability in the acknowledgment of a compliance with accepted standards of nursing care of the critically ill and a commitment to act within ethical principles. The critical care staff nurse is the one constant in the critical care setting because this role calls for a vigil at the bedside. As such, the critical care nurse is the coordinator of all care delivered by various health care professionals. The critical care nurse uses independent, dependent, and interdependent interventions in the management of the patient.

Examples of occasions when the staff nurse can create synergistic relationships might include the following: (a) serving as preceptor to a nursing student who opts for employment in that unit after an exceptional experience; (b) taking part in a unit committee and recommending changes in procedure or approaches to care using research findings reported in the literature, and (c) participating in a peer review performance evaluation system. These examples reflect situations that demand collaboration in order to yield positive results. In addition, they present opportunities for retention and recruitment of nurses, a responsibility of all nurses in the critical care unit.

Clinical Nurse Specialist

The clinical nurse specialist (CNS) role is multifaceted and includes the components of clinician, educator, consultant, researcher, and, to some degree, manager. The expertise and role of the clinical nurse specialist are derived from the combination of graduate study and clinical experience. The CNS draws upon a focused body of knowledge for making clinical decisions within a selected area of specialization in nursing (18). As the title implies, the practice of the CNS is more specifically focused than is the practice of a nurse generalist.

The CNS, as an expert clinician, may have a patient population determined by unit, service, or diagnostic grouping. The position of the CNS and the respective reporting relationships vary across institutions. For example, the critical care CNS may report to a nurse manager, a clinical director, or the Director of Nursing depending on the organization's decision-making philosophy.

Collaborative decision-making is imperative to the success of the role of the critical care CNS. The CNS collaborates with other nurses, physicians, patients, family members, and other care providers to facilitate efficient patient care and optimal patient outcomes based on technological, economical, social, and environmental factors. Some examples of the CNS engaging in productive collaboration include:

1. Acting in the role of study coordinator for multidisciplinary quality assurance study;
2. Acting as chairperson of the departmental standards committee;

5. Establishing and fostering the growth of a support group for neuro-trauma patients.

Nursing Case Manager

A role that is finding its place in acute care settings today is the nursing case manager. A case management practice model fosters an integrated approach to patient care whereby the coordination and continuity of care is facilitated by one individual called the case manager. The case manager provides leadership for the collaborative management of patient care and is accountable for the resulting patient outcomes. Depending on the institution, the case manager may be unit-based or service-based (clinical specialty).

The greatest distinction of this role is the fiscal management component that demands collaboration with physicians, administrators, and third party payors in order to achieve expected outcomes for cases within an appropriate length of stay and cost. The case manager defines a "critical pathway" or a day-by-day interdisciplinary plan of care that outlines the steps to be followed to achieve these outcomes. This time-lined plan of care epitomizes interdisciplinary collaboration as well as effective and efficient resource utilization. In contrast to the CNS, the role may not involve as much direct clinical practice.

Critical Care Nurse Manager

The critical care nurse manager coordinates specialty-based practice among many types and levels of caregivers through planning, organizing, leading, and controlling others. Performing these management processes requires a keen sense of balance and perception. The nurse manager is the administrative link for the staff nurse to higher levels of organizational decision-making, other patient care units, and ancillary services. Positioning of the role of the nurse manager is dependent on organizational size and decision-making philosophy.

The critical care nurse manager is not only a leader in unit-based and perhaps departmental and organizational decision-making, but is also a role model for staff members in clinical practice. Proficient clinical knowledge, skills, and competencies are important in maintaining critical care nursing standards.

The nurse manager is responsible for directing and evaluating the performance of the staff within a defined unit or set of units. In this function, the manager evaluates one aspect of the quality of care. Monitoring the quality of care and maintaining a unit-based quality assurance program are essential parts of the nurse manager role.

One of the most important roles the critical care nurse manager assumes is that of a team builder. The successful nurse manager collaborates with the staff regarding resource utilization, patient care issues, system issues, and delivery of patient services. By discussing issues with staff members and including their input into management decisions, the nurse manager creates an atmosphere that supports teamwork.

The manager who enhances the creative potential of a group and promotes synergistic relationships move beyond the delicate balance of managing and into behaviors such as leading, innovating, and creating. This creates growth for not only the staff but also the manager. Stevens (19) postulates that nursing leaders have particular skill associated with goal-setting, bridging communication, analyzing, problem-solving, and "negating" or ridding the environment of unnecessary bureaucratic procedures or outmoded practices.

Technical/Support and Assistive Personnel

With the explosion in critical care technology, stifling health care economics, increasing patient acuity and intensity of service, and the shortage of critical care nurses, there has been a proliferation of many types of technical, supportive, and/or assistive personnel in the critical care environment. For example, roles such as hemodynamic or cardiovascular technicians, patient care assistants, nursing assistants, telemetry or monitor technicians, and computer support technicians are becoming more visible in many critical care units. The manner in which these assistive roles are introduced has a great impact on their effectiveness. Well-defined roles, articulated levels of competency required within specific clinical areas, and clear lines of responsibility and accountability set the stage for mutual respect and collaborative efforts in the care of the critically ill patient.

In order to minimize confusion, frustration, fragmentation, and duplication of care at the bedside, it is important for the critical care nurse to maintain open lines of communication with all support personnel. It is within the scope of the nurse's role as the coordinator of the patient's care to establish the priorities of care, to define tasks, activities, and processes that require expertise, and to establish parameters that ensure accountability for delegated duties.

Physicians

Physicians caring for critically ill patients collaborate continuously throughout a patient's stay. The areas of interdependent practice present opportunities for collaboration and joint decision-making. Smoyak (20) summarized the conditions necessary to support collegial work relationships between nurses and physicians as: (*a*) mutual agreement on a goal; (*b*) equality in status and personal interactions; (*c*) a shared base of scientific and professional knowledge and complementary diversity in skills, expertise, and practice; and (*d*) mutual trust and respect for the other's competence.

It is important that physicians are aware of nursing routines and of the best times available for consultation and collaboration and vice versa, so that both can plan accordingly. Times can be set together for rounds, conferences, education, and patient care planning in order to provide the greatest likelihood of including all the appropriate team members. It is no longer acceptable for medical decisions to be made in a vacuum or for independent nursing interventions addressing new problems to occur without informing physicians. The approach to patient care must be collaborative and the results shared and evaluated on a continuous basis.

The importance of positive working relationships between nurses and physicians has been the focus of several studies over the last decade. The Magnet Hospital Study identified the nature of nurse-physician relationships as a key variable in nurses' job satisfaction (21). Nurses stated that they expected physicians to value nursing judgments and observations and to be consulted and involved in planning of patient care. Kerr (22) found that professional growth, development, and job satisfaction among the health care team members were positively correlated with collaborative and respectful work relationships. Two other studies also found that work relationships among nurses, physicians, and other team members affected the quality of care and consequently affected patient outcomes (23,24).

Other studies have demonstrated an inverse relationship between the nurse's level of participation in clinical and policy decision-making loops and nursing turnover rates (25, 26). The Secretary's Commission on Nursing recommended to employers of nurses and physicians that appropriate clinical decision-making authority of nurses in relationship to other health care professionals be recognized and fostered (27). Additionally, the Commission stressed the importance of close cooperation, mutual respect, and collaboration between nurses and physicians.

Expanded Nursing Roles

There are many nursing professionals whose practices are not confined to the critical care setting, but who consistently offer nursing expertise to critically ill patients and consultation to critical care nurses. The enterostomal therapist, infection control practitioner, staff development coordinator, quality assurance nurse, and discharge coordinator are examples of such roles. The presence and use of these roles is often dependent upon the practice model or system of patient care delivery and the manner in which nursing development and support programs are structured. The critical care nurse can facilitate the practice of these professionals by maintaining open channels of communication, incorporating their specific assessments and recommendations into the patient care plan, and inviting their participation in the evaluation of patient outcomes.

ROLE OF THE NURSE MANAGER IN PROMOTING A COLLABORATIVE ENVIRONMENT

The nurse manager sets the tone at the unit level and models effective collaboration through individual actions. Together with staff, the manager establishes guidelines and expectations for collaboration and healthy

.orking relationships within the organization. It is also the nurse manager who monitors and refines the system.

Intraunit Collaboration

The most important group on which the nurse manager focuses attention is the unit(s) that he/she directly supervises. Building internal collaborative processes is a key determinant of success in accomplishing goals and expected patient outcomes.

Leadership and Management Style

The manager must demonstrate leadership to the staff member. The manager establishes, with the input of the staff members, the model for delivery of nursing care. It is expected that this model is consistent and in harmony with other delivery systems within the nursing organization. For the most part, a single model is implemented throughout the organization and is part of the overall philosophy of the nursing service.

It is widely accepted that when nurses are given more responsibility and authority, greater job satisfaction, a reduction in absenteeism, and a decrease in turnover are the results (28). These factors are directly related to the quality of care and, as such, become the focus of management. Implementing a professional practice model is a way for the manager to encourage autonomy, to foster independent decision-making, to delegate authority as well as responsibility to staff nurses, and to establish an environment that is consistent with retention of nurses. A professional practice model has two major attributes: (*a*) a method of delivering care that vests responsibility and authority for patient care with the staff; and (*b*) decentralization of authority and decision-making to the staff nurse.

In a professional practice model, decisions are delegated to staff members for both patient care issues as well as daily unit management issues such as resource utilization (staffing), basic guidance of staff in interpersonal conflict, collaboration with other departments and individuals, and setting work priorities. The manager provides leadership to oversee the delegation of authority and responsibility for these decisions and is available for consultation and feedback to staff members as they learn how to fulfill these expanded components of their roles. The manager needs to encourage staff members to grow and participate in designing the future of the unit. This is accomplished by providing leadership for group problem-solving and decision implementation.

Team Building

Another important activity for the nurse manager is team building. Team building activities establish or strengthen cohesiveness among staff members, set group expectations and norms, and allow for a better understanding of the differences in the group. LeBella and Leach (29) outlined some fundamentals of facilitating group behavior listed in Table 12.2, which they call the "ABCs of team building."

The goals of team building are to establish a sense of belonging for all individuals, to monitor adherence to group norms and expectations of one another, to modify these norms as needed, and to enhance the environment for working together in order to build commitment and dedication to one another, the patients, and the unit.

Table 12.2. ABCs of Team Building

A. Assess
  Assess the circumstances and identify issues that will require attention; target goals and objectives to be achieved
B. Brainstorm
  Facilitate an open exchange of ideas and strategies for goal achievement, valuing diversity and differences in opinion being careful not to stifle the exchange by making judgments
C. Consider
  Actively listen and consider the merit of each suggestion by exploring each as a creative and innovative approach
D. Decide and Implement
  Select options that would be most likely to target the predetermined objectives merging differing roles so as to have a representative plan of action; use objective criteria in the selection process so as to minimize the sense of rejection by certain team members
E. Evaluate
  Monitor group behavior, the progress toward goal achievement, and be sensitive to needs of group reinforcement (29).

Interactions with Physicians

Nurse-physician collaboration and positive relationships are important in the critical care setting. Every critically ill patient requires both nursing and medical care. The nurse manager acts as an advocate for the staff nurse and ensures that opportunities and mechanisms for collaboration are in place. Further, the manager facilitates the nurse's having time or coverage to meet with physicians to discuss the plan of care, to provide input, to evaluate patient responses, and to recommend revisions to the plan of care. It is essential that this interaction and collaboration become incorporated as routine operating procedure in the critical care unit.

One way to ensure that this occurs is for the nurse manager to participate in orienting the rotating house staff members who come to the critical care unit. The nurse manager outlines the expectations for collaboration, for these will be reinforced by the physician director or other attending physicians. The nurse manager further provides oversight and, if necessary, enforces the established routines that are designed to yield the optimal patient outcomes as well as the most satisfying work environment for the staff nurses.

Interactions with Others Direct Care Givers

Individuals from many departments provide services to patients. Those who provide direct care include assistants and technicians within nursing and other technicians and professionals outside of nursing. The nurse manager is responsible for providing access to patients and families, while also protecting the patient and maintaining a therapeutic environment. The critical care nurse manager delegates responsibility and authority to staff members to coordinate all direct care that a patient receives. This includes authority to affect such things as the scheduling of tests outside the critical care unit, the review of the quality of care rendered by other individuals, and the changing of priorities on a moment's notice. The staff nurse is responsible for clearly communicating expectations of others, decisions about changes in the plan of care that might be disruptive to the other caregiver, and, most important, letting the patient and family know who these other individuals are and what interventions they provide.

Relationships with other caregivers in departments outside of nursing are crucial to completing a total plan of care. There is an interdependence among the services they provide. The nurse manager entrusts staff to communicate effectively and to cultivate the necessary relationships in order to maintain a collaborative environment.

Interaction With Indirect Caregivers

Some of the greatest frustrations that nurses experience are related to relationships with departments that provide indirect services. These include such departments as laundry, dietary services, central supply, pharmacy, and the laboratory. Services from these departments are essential. It is not uncommon, however, for nurses to feel as if not all these departments are working toward the same goal of positive patient outcomes.

Courteous and clear communication is a standard prerequisite to collaboration with these departments. In addition, the nurse manager must take the lead to invite individuals from other departments to become familiar with the critical care unit, its needs, both routine and emergent, the staff, patients, and expectations of the roles of these other departments in support of the mission of patient care. It is also important to listen to those other individuals to uncover what needs they have in order to be able to service the unit best. From that point, collaboration is the seeking of mutually agreeable solutions to interdepartmental problems. When communication breaks down or there is question or concern about the authority, responsibility, and accountability of other departments, the nurse manager as the unit's official representative to other departments in the organization seeks out the appropriate individual to solve the immediate conflict and to restore normal working relationships.

Intraorganization Representation

The role of the manager is best described by Likert's "linking pin" concept (30). This concept stresses how important it is for the manager to be able to exert influence upward in the organization in order to supervise successfully. Likert further explains that the linking pin role is executed differently depending on the maturity of the group. The greater the maturity level of the group, the more time the manager may spend in off-unit (linking pin) activities. For examples, in a critical care unit where the nurse manager has delegated to the staff the responsibility and authority for decision-making related to daily operations, the nurse in charge triages patient admissions, confers on discharges, troubleshoots supply problems, and communicates with physicians, hospital and nursing administrators, and others regarding the care of patients. The nurse manager is available for consultation and support and has given the staff latitude for decision-making and authority inasmuch as the manager does not have to be present on the unit continuously to provide oversight, nor does the manager have to be intimately involved in every decision or interaction in the daily operation of the unit.

Progressive critical care units with effective collaboration are usually noticed for their successes and accomplishments. Conversely, units with little or no collaboration are noticed for their lack of organizational participation and their lack of being team players, leading to a reduction of productivity, and an abundance of stagnation. All the units in an organization benefit from synergy where $1+1>2$. The critical care nurse manager, through promotion of collaboration and cohesiveness, can position a unit to contribute to a synergistic relationship and add to the productivity.

IMPACT OF RELATIONSHIPS ON UNIT WORK AND PATIENT OUTCOMES

The greater the loyalty of the individuals in a group, the greater the motivation among the members to achieve the goals of the group (31). When the nurse manager demonstrates supportive behavior and provides for group decision-making, there is an increase in loyalty of the staff. The energy of this loyalty is linked to positive results—in patient care, in job satisfaction, in meeting management goals, and in contributing to a synergistic relationship within the organization.

Good team work yields positive results. As a critical care unit's staff members work together, they build a sense of ownership of the unit. This ownership extends beyond the physical environment and, more important, includes such factors as the unit's decisions, its image and reputation, its representation on important committees in the organization, and its success in meeting major goals. Nursing staff and nurse managers work together to reduce absenteeism and turnover, to improve job satisfaction, to enhance growth opportunities, to develop leadership skills, to promote autonomy, and to create a more professional environment. Patient outcomes and the quality of patient care are directly affected by these factors. The more positive the environment, the better the outcomes. In a critical care unit where the staff members work together toward high productivity, improving job satisfaction, and creating an environment that is professionally rewarding, the result is a well-coordinated and cohesive team dedicated to the delivery of quality patient care. The group's loyalty and commitment are expressed in the work of improving patient outcomes and overcoming all obstacles in the path toward the end.

REFERENCES

1. Bolman LG, Deal TE. Modern Approaches to Understanding and Managing Organizations. San Francisco: Jossey-Bass, 1988: p 65.
2. Argyris C. Personality and Organization. New York: Harper and Row, 1957: p 49–50.
3. Argyris C. Integrating the Individual and the Organization. New York: John Wiley & Sons, 1964: p 7.
4. Scott WG, Mitchell TR, Birnbaum PH. Organization Theory: A Structural and Behavioral Analysis, 4th ed. Homewood, IL: Richard D. Irwin, 1981: p 102.
5. Gibson JL, Ivancevich JM, Donnelly JH. Organizations: Behavior, Structure, Process, 6th ed. Plano, TX: Business Publications, 1988: p 293.
6. Gibson JL, Ivancevich, JM, Donnelly JH. Organizations: Behavior, Structure, Process, 6th ed. Plano, TX: Business Publications, 1988: p 295.
7. Homans GC. The Human Group. New York: Harcourt, Brace & World, 1950: p 232–244.
8. Leavitt HJ. Managerial Psychology, 4th ed. Chicago: University of Chicago Press, 1978: p 357.
9. Maier N. Assets and liabilities in group problem solving. Psychological Review 1967;74:239–249.
10. Schein EH. Intergroup problems in organizations. In: French W, Bell C, Zawacki R (eds). Organization Development: Theory, Practice and Research, 2nd ed. Plano, TX: Business Publications, 1983: p 106–110.
11. Knaus, WA, Draper EA, Wagner DP, Zimmerman

JE. An evaluation of outcome from intensive care in major medical centers. Ann Intern Med 1986;104:416.
12. Koerner BL, Cohen JR, Armstrong, DM. Professional behavior in collaborative practice. J Nurs Adm 1986;16:42–43.
13. Guidelines for establishing joint or collaborative practice in hospitals. Chicago: National Joint Practice Commission, 1981: p 1.
14. National Commission on Nursing: Summary report and recommendations. Chicago: The Hospital Research and Educational Trust, 1983: p 26.
15. Joint Commission on Accreditation of Healthcare Organizations. Accreditation Manual for Hospitals, 1990. Chicago: JCAHO, 1989:251.
16. American Association of Critical Care Nurses. Collaborative Practice Model: The Organization of Human Resources in Critical Care Units. Newport Beach, CA: AACN, 1982.
17. Conner DR, Palmgren CL. Building Synergistic Work Teams to Cope with Organizational Change. Atlanta: Organization Development Resources Press, 1980: p 2–6.
18. American Nurses Association. The role of the clinical nurse specialist. Kansas City, MO: ANA, 1986: p 1.
19. Stevens BJ. The Nurse as Executive, 3rd ed. Rockville, MD: Aspen Publishers, 1985: p 238.
20. Smoyak SA. Problems in interprofessional relations. Bull NY Acad Med 1977;53:51–59.
21. McClure ML, Poulin MA, Sovie MD, Wandelt MA. Magnet Hospitals: Attraction and Retention of Professional Nurses. Kansas City, MO: American Nurses Association, 1982: p 32.
22. Kerr J. Interpersonal distance of hospital staff. West J Nurs Res 1986;8:350–364.
23. Knaus WA, Draper EA, Wagner DP, Zimmerman JE. An evaluation of outcome from intensive care in major medical centers. Ann Intern Med 1986; 104:416.
24. Rubenstein LZ, Josephson KR, Wieland GD, English PA, Sayre JZ, Kane RL. Effectiveness of a geriatric evaluation unit. N Engl J Med 1984;311:1664–1670.
25. Price J, Mueller C. Professional Turnover: The Case of Nurses. New York: Spectrum, 1981: p 55.
26. Weisman C, Alexander CS, Chase GA. Job satisfaction among hospital nurses: a longitudinal study. Health Serv Res 1989;15.
27. Secretary's Commission on Nursing. Final report, volume 1. Washington, DC: Department of Health and Human Services, 1988: p 34.
28. Simpson K, Sears R. Authority and responsibility delegation predicts quality of care. J Adv Nurs 1985;10:345.
29. LaBella A, Leach D. Personal Power. Boulder, CO: Newview Press, 1983: p 134–135.
30. Likert R. New Patterns of Management. New York: McGraw-Hill, 1961: p 14.
31. Hersey P, Blanchard K. Management or Organizational Behavior, 2nd ed. Englewood Cliffs, NJ: Prentice Hall, 1972: p 64.

Chapter 13
Customer Relations: A Supportive Environment for the Family

JOAN GYGAX SPICER, MARYANNE ROBINSON

Patients and their families may have perceptions that the critical care environment may present threats to the family's well-being. Unless these perceptions are addressed appropriately, adverse effects on patient care and to the family unit may result. Integrating a customer relations focus into the daily operations of the critical care environment is one method of creating a supportive environment for families during the time a family member is a patient in the critical care unit.

IDENTIFYING FAMILY NEEDS

A supportive environment can be described as a climate that promotes and maintains a hopeful state of well-being. Creating a supportive environment requires needs assessment and thoughtful interventions. A family's reaction to the crisis of critical illness or trauma can often give clues to underlying needs. These needs, in turn, are a basis for planning strategies in order to create a supportive environment for the family.

Figure 13.1 delineates the conceptual process of identifying family needs. In this model, the critical illness or trauma of a family member acts as a stressor on the family's well-being, which can manifest reactions at three levels: emotional, physical, and sociocultural. At the emotional level, reactions are human sensations provoked by the urgency of the crisis and often elicit such feelings as fear, anxiety, anger, and guilt. Corresponding needs of the family at this level include assurance and confidence that the patient is receiving adequate care and that the staff is competent and efficient. Bouman [1] identified in her study related to priority needs of the family that the family has an important need to know that the best care possible is being given to their family member. Other needs identified in this study that can be related to the emotional level include:

1. To know the probable outcome;
2. To know exactly what is being done for the patient;
3. To know how the patient is being treated medically;
4. To feel that hospital personnel care about the patient;
5. To feel there is hope [1].

At the physical level, reactions are often responses caused by sensory overload provoked by environmental stimulation. Feelings of being overwhelmed, closed-in, invaded, and powerless are reactions related to the physical level. Restricted visitation, limited waiting room facilities, and the "high technology" patient care unit may provoke such feelings. At this level, the needs of the family may include privacy, space, comfort, and instructions related to the patient care unit and visitation. Bouman's list of identified family needs that correspond to the physical level include:

1. To receive a telephone call at home about changes in the family member's condition;
2. To see the patient frequently;
3. To have explanations about the environment before going into the intensive care unit [1].

Molter [2] outlined other needs related to the physical level including:

1. To have the waiting room near the patient;
2. To have a bathroom near the waiting room;
3. To have comfortable furniture in the waiting room;

Molter identifies other needs which may related to the sociocultural level. These include:

1. To talk to the doctor at least once a day;
2. To be told about other people in the hospital who could help the family member;
3. To have the pastor visit (2).

The critical care nurse is key to identifying family needs. The nurse's responsibilities include managing and controlling elements of the environment which may threaten the stability of the family unit and thereby impact patient care. Figure 13.2 describes the relationship of the family to the critical care environment.

A CUSTOMER RELATIONS FOCUS

The focus of customer relations is client satisfaction and service integrity. Nursing is a service, part of which is provided to the family. A service by its very nature is experiential, and the providers of the service need to ensure its integrity or completeness as it is perceived by the family (3). Creating a supportive environment focuses on three principles of customer relations: interpersonal interaction, physical and policy manipulation, and resource development.

Interpersonal Interaction

In the critical care environment, families look for tangible cues that correspond with their perceptions of quality. Tangible cues can be a smile, a tone of voice, eye contact, and successful problem solving. Staff members who are perceived as caring and efficient can minimize the stress the family may feel as well as project the quality of care being provided to the patient.

Open communication is perhaps the most crucial element in creating a supportive environment and meeting the needs of the family. One of the important roles of the nurse manager in creating a supportive environment is to facilitate open communication between nursing staff members, and the patient and family.

The major objective involved in meeting the needs of the emotional level by incorporating human interaction principles is the promotion of strategies that build up the family's trust

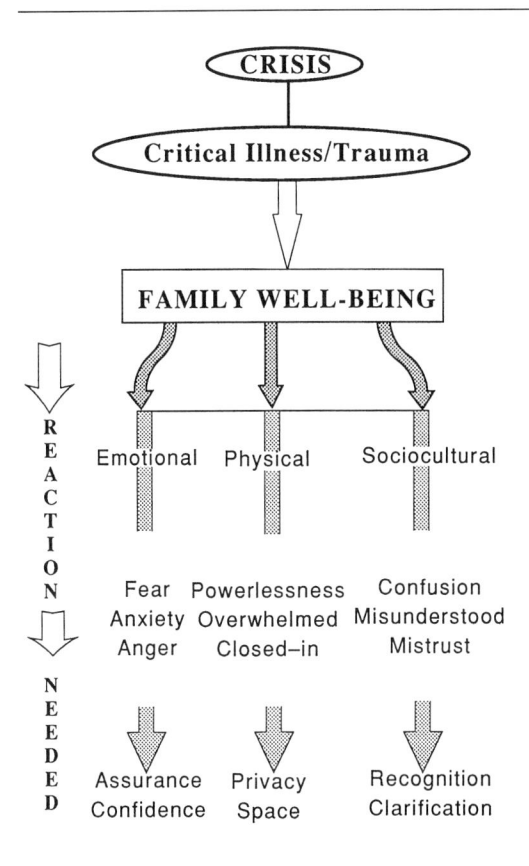

Figure 13.1. Family needs.

4. To have a telephone near the waiting room (2).

Sociocultural reactions stem from internal conflicts with personal values or beliefs. Feelings of confusion, being misunderstood, hostility, and mistrust are common examples of such reactions. Differences in ethnic values may provoke feelings of misunderstanding and mistrust of medical care. Needs at this level include recognition, clarification, and knowledge of what is "going on" with the patient. Family needs at this level may include:

1. To know specific facts about the patient's program;
2. To have questions answered honestly;
3. To talk about the possibility of the patient's death (1).

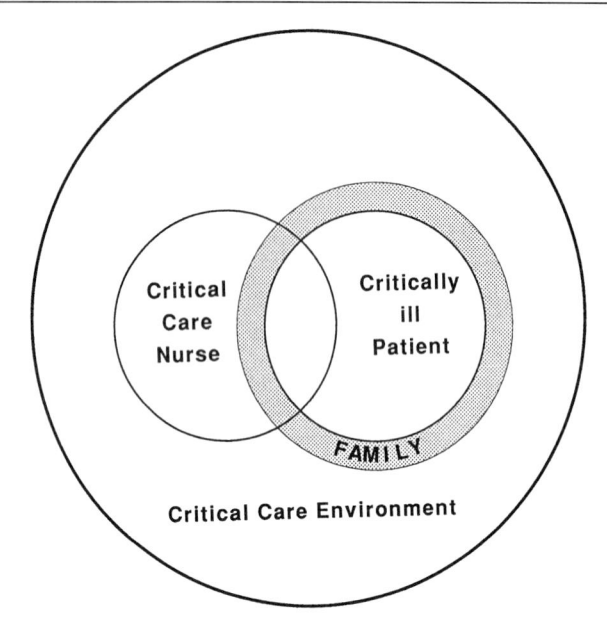

Figure 13.2. Family unit relationship to the critical care environment.

level and provide evidence of quality assurance. Some strategies include:

1. Staff training on "guest relations" concepts, inclusion of behavior expectation on performance evaluation;
2. Providing opportunities for patient care conferences for the family to include appropriate members of the health care team;
3. Promoting the family spokesperson role;
4. Developing a delivery care model that promotes a case manager or primary nurse;
5. Establishing minimum criteria for frequency and content of nursing staff-family interactions.

Physical and Policy Manipulation

Family needs with respect to the physical environment include what Carpman et al. (4) defined as wayfinding or the ease in which families can find their way around the building. The ease or lack of ease of wayfinding activity has a direct impact on the level of stress that families already feel. Entry points into the hospital should put the family in visual alignment with a reception or information area. Families need to know where they need to go. A wayfinding system includes signs, color coding, and information. Messages on signs must be at the appropriate level for the targeted audience. Research identifies that for the general audience this is somewhere between a sixth and eight grade reading and vocabulary level. Because signs are utilized by more than one audience, such as care providers, a decision has to be made as to which audience to target by signs to provide consistency in signage. In areas to which patients, families, and visitors have access, signage should be consistently targeted to these audiences.

When under stress and preoccupied, family members often perceive the corridors as tangled and as leading into nondirectional pathways. The family members frequently cannot find their ways through the hallways to the critical care unit. Even if they have been there once before, they cannot necessarily rely on that experience. "You-are-here" maps placed along the corridors can help support the family under stress.

The wayfinding is usually aimed toward the family waiting area. The physical setting and amenities in the waiting area are dependent on the organization's space and financial resources. There are some considerations, both physically and in the management of the waiting area, that have to be dealt with, since the critical care manager may also have responsibility for managing the family waiting area.

Molter (2) identifies the family's need to have a waiting area near the patient. Not all physical plants can accommodate this, so there may be times during the critically ill patient's hospitalization that temporary arrangements need to be made to facilitate the family's being physically close to the critical care unit. Also, Carpman et al. (4) identified the following physical amenities as important in the waiting area: clocks, telephones, restrooms, drinking fountains, and diversionary items, such as reading material, and television. Another consideration in physical manipulation is noise and traffic control. What the family hears and observes are tangible cues associated with their evaluation of the quality of care given to the patient. The family waiting area needs to be out of traffic patterns of employees and physicians and out of hearing distance of employee conversations.

The physical size, location, and furnishings of the waiting area may lend to the management of the waiting area by the families; however, when the waiting area is physically decentralized, nurse managers have to remain cognizant that they may have to intervene and manage not only the waiting area but also the waiting families.

While waiting families need to know that they have not been forgotten by those in control of the patient, a standard should be established by the nursing staff as to frequency and content of family communications. Families need to know what is expected of them and what they need to do to meet these expectations. A common form of communicating these expectations is a written visitor's information guide outlining visitors' expectations and what visitors can expect of the care providers. Appendix C is an example of a visitors' information guide.

Families need to have choices as to whether they have to interact with other patient's families. There may be times when there are incompatibilities within the family unit or between families. Monitoring of the interpersonal interactions is necessary to prevent unneeded tension in an already stressful situation.

Among other recommendations related to physical and policy manipulation are:

1. Developing and initiating contracts with family as appropriate;
2. Utilizing a concierge program;
3. Designing seating arrangements to maximize privacy and space;
4. Using "quiet rooms" for grieving families and conferences;
5. Developing diversionary media, such as educational videotapes, informational booklets; and
6. Implementing flexible visiting rules.

The primary objective of management strategies in physical and policy manipulation is to minimize the additional stress provoked by sensory stimulations. Strategies within this level are concerned with control of space, distance, and comfort.

Resource Development

The last principle related to customer relations is resource development in support of the family. Resource development is associated with the sociocultural level of family needs. This is the most complex of the levels because of the basic assumption that these needs are created by interpersonal conflict, such as ethnic values and socioeconomic levels. Although many of the strategies mentioned in human interaction and physical manipulation can be used to meet the needs springing from the sociocultural level, the major objective at this level is to increase the family's awareness of what is "going on" with the patient and establishing and maintaining a trust level between the care provider and the family. This assists family members in developing their individual and/or collective coping mechanisms in accepting the patient's illness.

Management strategies include the following:

1. Integrating other services, particularly social service and spiritual support;
2. Developing support groups;
3. Developing educational media for family and patient education programs;
4. Developing preoperative educational programs to include tours of the critical care unit for patients scheduled for elective procedures with potential admission to critical care i.e., open heart surgery;

5. Increasing awareness of both internal and external referral resources for future needs of the patient, particularly the identification of resources related to cultural and ethnic backgrounds for family counseling;
6. Utilizing interpreter services for communication.

An additional objective that is not addressed in this discussion, but is important to mention here, is the need to evaluate systems that directly affect the nurse. If the systems are not supportive of the care provider and require time and attention to maintain, the nurse may become desensitized to the family because energy is being directed to attempt to obtain services from poor or nonfunctioning systems for the patient.

PATIENT/FAMILY FEEDBACK AND EVALUATION OF SERVICE

The nurse manager is responsible for providing a method for soliciting feedback and evaluation of service from the patient and/or the family. One of the traditional methods of obtaining feedback is by posthospital discharge survey.

Research has shown that patients' overall satisfaction with health care services and their intention to revisit the providing organizations are directly influenced by both perceptions of the organizations' performance and the fulfillment of performance expectations and perceived fairness (5). Posthospitalization surveys need to incorporate measures of general service satisfaction, perceptions of fulfillments of expectations and organizational fairness, overall satisfaction, and intention to visit the organization for future service (6).

Interviewing the patient/family posttransfer from the critical care unit might be more appropriate than from postdischarge from the hospital. During the interview, information can be elicited from patients and their families about satisfaction with the nursing care and with the support received by the family. If questions remain unanswered about the patient/family's perceptions concerning the experience in the critical care unit, these feelings can be further probed at this time. There is an opportunity to decrease dissonance if the perceived experience of the service was not what was expected by the patient/family. The post-transfer interview can be thought of as analogous to service call back. Service call back is also an extension of the service.

A procedure for responding to patient and family feedback needs to outline a timeframe for response, who will respond, and how the response will be made. If the feedback reflects a negative evaluation, a standard for response may be that within 24 hours after the nurse manager is aware of the patient or family concern, a telephone follow-up is initiated by the nurse manager and a written response is made within 5 working days. A written response should include a statement of the understanding of the concern, outline of what is being done about the concern, specific mention of the use of the patient/family recommendations in resolving the concern, and a statement of the desire for the patient's/family's satisfactory experience the next time the service is utilized (6).

CONCLUSION

Molter (2) cited three assumptions that should be considered when integrating the family into the concept of quality of care. Family participation is essential to provide quality care. All individuals within the human relations sphere of the environment are products of some family unit. Finally, the family units remains a viable way of life for the majority of the population (2).

In Appendix D, Bouman (1) outlined helpful interventions and strategies related to caring for the family of the critically ill patient. This family-client care plan provides an excellent resource for nurses when addressing family needs.

The family support structure is vital to the care of the patient. Caring for the critical care patient entails a commitment to supporting the family. The nurse manager has a major responsibility to promote and facilitate a supportive environment. This is possible by creating an environment focused on family needs and by implementing a customer relations program. A customer relations focus includes: review of the physical setting for appropriateness in supporting the family, structuring frequency and content of interactions with families, and providing feedback mechanisms for evaluation of service.

REFERENCES

1. Bouman CC. Identifying priority concerns of families of ICU patients. Dimens Crit Care Nurs 1984;3(5):313–319.
2. Molter NC. Needs of relatives of critically ill patients: A descriptive study. Heart Lung 1979;8(2):332–339.
3. Spicer JG, Craft MJ, Ross K. A systems approach to customer satisfaction. Nurs Adm Q 1988;12(3):79–83.
4. Carpman JR, Grant MA, Simmons DA. Designs That Care. Chicago: American Hospital Association, 1986.
5. Swan JE, et al. Deepening the understanding of hospital patient satisfaction: fulfillment and equity effects. Journal of Health Care Marketing 1985(Summer)1985:7.
6. Spicer JG, Mickley RB. The nurse as a marketer. In Lewis EM, Spicer JG (eds). Human Resource Management Handbook. Rockville, MD: Aspen Publishers, 1987:p 89–98.

Chapter 14
The Critical Care Patient: Characteristics and External Environmental Influences

KATHLEEN E. ELLSTROM, MARGARET MACARI-HINSON

Patients are admitted to the critical care unit because of an actual or potential life-threatening event. However, the climate of health care today is so dynamic and changing so rapidly that factors external to the critical care environment play major roles in determining the characteristics of the patient requiring critical care services. Therefore, it is beneficial for administrators and managers to assess and evaluate these factors to enable them to meet the health care demands of their patients better.

SURVIVABILITY FACTORS

The survivability of patients who have sustained life-threatening illness or trauma has increased because of the development and implementation of prehospital care programs. These programs include training programs for professional and public community members on cardiac life support, community emergency medical systems (EMS), and enhanced accessibility to emergency health care services.

The single major factor that has had an impact on the characteristic of today's critical care patient is the wide use of cardiac life support techniques as well as the training of community and professional people on these techniques. Since the National Conference on Standards of Cardiopulmonary Resuscitation (CPR) and Emergency Cardiac Care (ECC) in 1973 made recommendations regarding recognition of early warning signs of heart attack, CPR training, and advanced cardiac life support, research has shown survival from cardiac arrest due to ventricular fibrillation has increased dramatically (1).

An emergency medical system (EMS) can be described as a comprehensive, community-coordinated program providing ready access to life-saving health care services through an integrated network of professional resources. Hunt's review of the literature cites factors of EMS programs associated with higher admissions and discharge rates. These factors include bystander CPR, initiation of CPR within 4 minutes, paramedic rescue, and initiation of advanced cardiac life support within 10 minutes (2). In addition, community involvement, community and professional training, and integrated communication systems were among other factors associated with successful EMS programs that were related to the survival of sudden cardiac victims.

Trauma victims have also benefited from the EMS program. In the last decade, designated trauma centers emerged nationwide in a response to meeting the needs of victims sustaining major trauma, such as motor vehicle accidents. The first hour after a trauma is the most crucial period. Baxt and Moody (3) noted a significant decline in the mortality rate when trauma patients receive rapid, definitive care within the first hour, correlating with higher positive outcomes. Overall, CPR training, implementation of the EMS, and the emergence of trauma centers have collectively saved lives which otherwise would have been lost. The significance of these factors has influenced the intensity of services required for these survivors in the critical care units.

The benefits for trauma patients of rapid access to a systematic, organized approach was further illustrated by Moylan et al. (4) with their evaluation of 330 trauma victims. A comparison of air versus ground transportation revealed a significant increase in survival for air-transported patients. This difference may be attributed to the fact that those patients receiving air transportation received more interventions (e.g., blood, intubation, military antishock trouser [MAST] suit) than those transported by ground. Vital signs of air-transported patients remained stable (4).

Although medical costs have increased with the continuing use of rapid transport systems, i.e., helicopter service for transporting critically ill patients from the scene to the hospital or hospital-to-hospital, patient outcomes have generally been positive. One study examined the effect of mobile paramedic units on outcomes of patients with an acute myocardial infarction (MI). The study compared 134 patients who received care for MI from the paramedics with 101 who did not receive care. They found that not only was the outcome not improved by using paramedic care, but that, additionally, there was a 29-minute median delay in the transport of the patient. The only beneficial treatment that paramedics could give was defibrillation, and that was offset by the delay in getting to the hospital (5).

Another study evaluated patients with penetrating cardiac trauma over a 2-year period. One factor associated with mortality was the mode of transportation to the hospital. The conclusion of the study was the rapid transport, aggressive resuscitation, and cardiorrhaphy are the best treatment for penetrating cardiac trauma (6).

One recent review of patients with cardiac arrest secondary to trauma and the impact of aggressive physician intervention at the scene as well as rapid air transport to a trauma center showed that physician intervention on the scene and rapid transport did not improve the outcome in this type of injury (7). A review of the autopsy data showed, however, that the majority of patients in this study had major injuries that were incompatible with life.

The role of helicopter transport in rural areas was evaluated by Urdaneta et al. (8). Records of 916 trauma patients transported by helicopter were reviewed for impact on outcome. In only 27% of the cases was helicopter transport considered helpful or essential, yet, because it is virtually impossible to determine which patients hypothetically would benefit from air transport, the authors concluded that its use is beneficial especially in rural areas (8).

Two studies evaluated aeromedical transport services using the TISS (Therapeutic Intervention Scoring System). One study looked at five different programs in order to evaluate whether patients being transported in this manner actually needed that level of intensive transport. With an average of 41.5% of the flights appropriate according to the index and another 35% having greater than 50% probability of being appropriate, the authors concluded that the flights were appropriate most of the time (9). The other group evaluated the impact of a physician-accompanied transport team. In 23% of the patients, significant interventions such as intubation, mechanical ventilation, or continuous vasopressor support were necessary before departing from the referring hospital. The addition of a physician to the team offered advantages in the treatment available for morbidity over long distances and prolonged periods of transport time (10).

Overall, because of the impact on patient outcome, the decrease in transport time in order to make aggressive therapy available for cardiac, trauma, or other injuries appears to compensate for the costs involved in training and maintaining the aeromedical teams.

SOCIOECONOMIC FACTORS

The socioeconomic factors of the health care industry have been primary forces influencing the characteristics of the critical care patient population. The implementation of Diagnostic Related Groups (DRGs) in 1983 as the mechanism for a prospective payment system (PPS) has forced the health care market to change. Needless to say, the provision of health care is not the same as it was 20 years ago.

The major issues facing health care professionals related to PPS are the efficacy of measuring cost-containment strategies and their effects on technology affecting the quality of care. In one study, Rettig et al. (11) cited the reduction in average length of stay (ALOS), admissions and occupancy rates from 1983–1984, and the fact that hospitals have generally increased profitability with PPS. Ahmad et al. (12) investigated the financial impact of DRGs for Medicare patients in the medical intensive care unit (MICU) comparing costs for 1984 and 1985 discharges. Although there were no changes in mortality rates or DRG categories, DRG payments covered 55% of the 1984 costs but only 46% of the 1985 costs. These authors suggest more studies are needed to measure appropriate utilization of the MICU and government subsidies based upon severity of illness.

In a study investigating Medicare PPS and technology, Sloan et al. (13) also found that PPS did not decrease the use of the intensive care unit (ICU) for Medicare and non-Medicare patients. The number of routine tests per patient, however, declined markedly for both groups after PPS was initiated.

Mushlin et al. (14) reported their experience during a 5-year period in which a community-wide experiment evaluated the effect of PPS on quality of life and outcome. Despite an increase in the number of admissions for maternal illness and acute myocardial infarction, the number of elective surgical procedures declines along with stable or improved outcomes in neonatal deaths, ischemic heart disease deaths, deaths from five selected surgical procedures, and rates of adverse outcome for medical and surgical conditions (14).

DRGs were originally developed to utilize an ALOS, and were later modified to reflect age and death. DesHarnais and associates (15) evaluated and resource utilization in patients older than 70 years during pre- and post-PPS time periods. Results indicated that although age accounted for minor variances in ALOS, it was patients with comorbidities and complications who accounted for the largest differences regardless of age. Although statistically significant due to the large sample size, the differences were not felt to be practically significant. These trends were maintained with age breakdown in 5-year categories. The authors concluded that increased age without comorbidities has little effect on ALOS and perhaps a revision in DRG categories is needed to produce more similar groups and equitable reimbursements (15).

Calore and Iezzone (16) evaluated the variability within DRGs due to illness severity by investigating the effects of disease staging and patient management categories (PMCs). The investigators concluded that DRGs were most efficient in explaining variation of cost; however, disease staging and PMCs may be helpful adjuncts to DRGs by refining unrelated comorbidity and by increasing equitable reimbursement. These systems may also identify a class of patients within a DRG category that does not belong because of unique resource needs (16).

In an analysis of 2431 medical Medicare patients to assess DRG stratification and resource consumption, Munoz et al. (17) assigned patients to one of four groups: patients without complication/comorbidity, patients with one or two complications/comorbidities, patients with three to four complications/comorbidities, and patients with greater than four complications/comorbidities. Analyses based upon ALOS, number of diagnoses, number of procedures, and profit/loss were conducted to assess resource consumption. Patients with greater than four complications/comorbidities had the highest resource consumption. The authors concluded that these financial characteristics may affect access to care and quality of care for this group of patients (17).

Sorrentino (18) reported on ALOS, charges, reimbursements, and death rates for 56 acute care JCAHO hospitals with over 100 beds. Although there were variations among non-profit, for-profit, and government hospitals for the 20 DRG categories analyzed, the evidence was not convincing in suggesting that for-profit hospitals limit ALOS. For-profit hospitals demonstrated higher charges and higher reimbursements. Analyses of death rates demonstrated no significant difference among the different hospitals and, therefore, no difference in the quality of care (18). Freund and associates (19) analyzed ALOS between investor-owned and voluntary hospitals. When case-mix variation and other hospital level factors (e.g., number of beds, occupancy rates, etc.) were controlled, there was no difference in ALOS. Investor-owned hospitals had an ALOS of 1.3 days fewer than the voluntary hospitals, but this was thought to be related to the hospitals actually having shorter stay patients (19).

In a 3-year analysis of over 14,000 pediatric patients, Munoz et al. (20) found that black and Hispanic patients had significantly more ICD-9-CM codes when compared to white patients. When adjusted for DRG weight index, case-mix, profit/loss, and emergency admissions, blacks and Hispanics had longer ALOS, higher hospital cost, and a greater number of ICD-9-CM codes per patient when compared to whites. Consequently, although there was under-reimbursement for all patients, there was significant under-reimbursement for blacks and Hispanics with the greatest percentage of outliers being blacks, who also had the highest emergency admission rate (20).

The concept of shorter stay has raised the question of sicker patients being hospitalized with a higher acuity, necessitating a higher intensity of care, but reduced availability of resources. Corcoran and Diers (21) evaluated nursing resources consumption for 102 cardiac surgery patients utilizing various tools to measure illness severity and patient acuity. There were no relationships between illness severity and total nursing resource consumption for ICU or patient acuity and routine floor care. The authors, therefore, raise questions about the duration of some ICU stays and the need for nursing to demonstrate its contribution to patient outcomes (21). Trofino (22), in her analysis of nursing care hours of over 10,000 cases in 48 DRG categories, found that nursing care has been affected by DRGs and ALOS. Nursing is attempting to comply with the ALOS constraints placed by DRGs. McCloskey (23) proposed a mechanism to determine costs for nursing services and charging them to third party payers. One suggestion was to multiply the nursing time for DRG intensity level by the nurse's salary and add this to the indirect costs. This model, however, does not measure nursing activities. Another suggestion was to utilize nursing diagnoses to generate nursing cost in DRG categories based upon nurse and physician-initiated treatment/interventions (23).

DRGs have reduced reimbursements to hospitals while containing costs. It is evident from the literature that more research is warranted to quantify cost-effectiveness of nursing services, particularly in critical care, to substantiate the characteristics and needs of the critical care patient better.

Another important factor is the medically indigent population, which has also had an impact on the characteristics of the critical care patient. The impact of the medically indigent patient on the health care system is fast approaching a crisis state. The medically indigent patients are those who are underinsured or not insured. Approximately 35 million Americans are uninsured for at least part of each year, and in any given time, about 25 million are uninsured (24). Medicare and Medicaid, as well as other state or federal programs, provide insufficient reimbursement for costs incurred during an acute illness. Only about 44% of the medical costs for the elderly are covered by Medicare. The percent of poor people receiving Medicaid has declined from 63% in 1975 to 46% in 1983 (24).

The noninsured health consumer is fast becoming a large segment of the population. Increasing costs associated with insurance and inability to obtain insurance contribute to noninsured population. Moreover, in many areas of the Southwest and Southeast, a growing segment of the medically indigent population is comprised of illegal aliens whose health care resources are very limited.

The impact on the critical care of the medically indigent and their lack of access to services is twofold: impact on resources (cost, equipment, personnel, etc.) and impact on patient acuity. Decreasing reimbursement for hospital costs has reduced the funds available for maintaining the function of the facility. The medically indigent patients who are admitted to critical care units are generally sicker and have a higher acuity than those who have ready access to health care services, particularly to preventive health care services. They frequently have chronic treatable conditions that are out of control because of inability to obtain or pay for medications—e.g., a diabetic admitted to the ICU in ketoacidosis because of lack of money to buy insulin. Often, multiple conditions are present and have been untreated, thus leading to multiple system organ failure (MSOF) and the associated higher morbidity and mortality.

POPULATION DEMOGRAPHIC FACTORS

The most significant factor in terms of impact on health care services is population demographics. As a result of advances in technology and medicine, life expectancy has increased. The aging population, coupled with the rising birth rate, has caused a concern related to future implications on health care.

The "graying" of America is providing new dilemmas in critical care. Many elderly are functioning well into their eighth and ninth decades, yet when a catastrophic illness occurs, the availability of resources is limited. The impact of critical care and acute illness on the elderly ranges from financial to social to physiological. The majority of elderly live on fixed incomes that do not allow for major expenses, and Medicare, in most cases, is the primary payor for medical expenses.

The increasing longevity of the population is resulting in patients with chronic health problems presenting themselves for surgery. With the elderly, there is an added weighing of the risks along with the benefits of surgery. Not only do the elderly present an increase in chronic health problems, but also they frequently assume that a symptom is part of the process of aging, when it could be a correctable problem. Many elderly grew up during a time when hospitals were places people went to die, and they then have a fear of dying and may be convinced that they are dying when actually the problem could be corrected. These beliefs often interfere with the ability of the health care team to provide health care to the elderly. The family support system of the elderly has also had an impact on the care of the elderly, because many times family members may not live nearby. Moreover, friends may not be able to visit because of a lack of transportation or their own physical disabilities.

Physiologically, the "slowing down" of the body affects all systems: cardiovascular, pulmonary, gastrointestinal, hormonal, etc. A decrease in cardiac output because of a decrease in muscle tone and relaxation of the smooth muscle has an effect on all systems: decreased mentation, decreased kidney output, decreased gas exchange in the lungs, etc. (25). The immune response of the elderly is also compromised, which limits their ability to deal with stress. They are more susceptible to infection, yet they do not respond as quickly to deal with the infection when it is evolving. The chronic nature of many of their problems tends to utilize their physiological resources so that one more event may be the trigger for catastrophe, and recovery may be prolonged or impossible. Because of this natural biological aging process, the elderly population generally requires more resources when they are admitted to the ICU. A study examining the relationship between cost of ICU care and outcome of illness to patient age determined that the cost of care is the same for younger patients as for the older patients (26). The notion of care rationing and resource allocation has been a major consideration (27). Two studies evaluating the response of patients over 65 years of age to critical care treatment showed that mortality cannot be predicted by advanced age alone. Even patients 85 years of age and older had reasonable outcomes from critical care. One exception was in the septic category, which showed an increase in mortality in the elderly (28).

Another contributing factor of the population demographics is the long-term needs of the many people surviving illnesses and traumas. The chronic nature of health problems in the population because of advances in technology and particularly the lengthening of the life-span have resulted in an increase in organ system failures when these patients are admitted to the ICU. The mortality rate of critically ill patients rises dramatically with the failure of more than one system (29). A recent study analyzing MSOF in trauma and nontrauma patients showed a mortality rate of 13% for one system failure; 34.5% for two systems; 75% for three systems; and 92.9% for four or more systems. Mortality was slightly increased with the elderly population (30). Another review places mortality at 98% for patients with three or more system failures persisting after 3 days. Biological age (chronological age and health status) is associated with an increase in mortality, rather than chronological age alone (31). The most common systems to fail are respiratory, cardiovascular, and nervous systems with sepsis increasing mortality significantly when associated with more than one system failing (30,32,33). In the pediatric population, however, sepsis does not appear to be as much of a factor in increasing mortality in the presence of MSOF (34). MSOF in transplant patients also is devastating and has a high mortality, which is caused by the patient's immunosuppressed state (35).

Various therapies are being explored to

reduce the incidence of major complications that lead to organ system failures. Continuous arteriovenous hemofiltration (CAVH) and continuous arteriovenous hemodialysis (CAVHD) are being used with good results in the presence of renal failure (36–38). Vasodilator prostaglandins have been used to increase oxygen delivery in patients with MSOF associated with sepsis (39). As sepsis has been implicated as a major determinant of MSOF and is frequently the last blow to an already overwhelmed critically ill patient, research is also being performed in the areas of antibiotic therapy and anti-inflammatory agents. (31,40). Another area of research is focused on the damage to the pulmonary system and on therapies or interventions that can prevent the damage to the pulmonary vasculature (41). A promising therapy is the use of a combined artificial lung and kidney to treat adult respiratory distress syndrome (42), although results of a random prospective study are not conclusive at this time (43).

On the opposite end of the life continuum, neonates are surviving at a younger gestational age. Infant mortality rates have steadily declined. In 1979, however, the United States ranked only 16th in the world in infant mortality. This may be related to improved technology in delivering very low birth weight (VLBW) infants as well as improved perinatal care, therefore resulting in the delivery of sicker infants. Evaluation of neonatal mortality (birth to 28 days) and postnatal mortality (28 days to 1 year) based upon black/white ratio indicated an increase in neonatal mortality while there was a decrease in postnatal mortality. This suggests that prenatal and perinatal problems reflect neonatal mortality, and environmental factors reflect postnatal mortality (44).

Resnick and colleagues (45) investigated the effect of birth weight, race, and sex on survival of over 16,000 low birth weight infants (500–2500 gm at birth) during a 7-year period. Survival was defined as attaining transfer to a step-down unit or discharge to the referring hospital after the neonatal intensive care (NICU) experience. Data were analyzed by 50-gm increments to provide more precise statistics. The effects of birth weight on survival were significant from 500–1500 gm.

Female infants had better survival rates than male infants in this weight group. Survival of white infants over black infants was greater in the 500 to 650-gm weight group; however, black infants demonstrated improved survival beginning at the 650 gm weight class. Male/female differences were the same for blacks/whites over all weight classes. Infants with birth weights of 1500–2500 gm were not significantly different among the variables and achieved nearly 95% survival. Survival rates did not differ among infants born at the hospital when compared to those transported from a referring facility. The authors suggest a more detailed description for infants in the 500 to 1500-gm weight category (45).

Contrary to the benign effects of transportation on survival cited by Resnick et al. (45), the study performed by Saule and associates (46) in their evaluation of the effectiveness of neonatal transport in West Germany concluded that the benefit of transport was limited. They suggested that in predicted high-risk cases (e.g., birth before 33 weeks' gestation) the mother should be transported before birth (46).

In her review article, McCormick (47) evaluated the long-term outcomes of VLBW infants (less than 1500 gm) discharged from NICUs. Analysis must be based upon the children's status as they enter school, providing a more precise indication of their ability to function as they approach adulthood. Discharge from the NICU may not be the end point for VLBW infants (47). The impact of these factors on these children and on their progress mentally and physically has yet to be fully established over their life-spans.

RESPONSES TO THE ENVIRONMENT

The environment of the critical care unit affects patients in many ways. Sensory overload and deprivation have many physiological and psychological effects on patients that can prolong the length of stay in the ICU or hospital as well as contribute to increased morbidity. The changing character of the critical patient with multiple chronic problems, immunosuppression, or of an age at either extreme of the life continuum places the patient at risk for iatrogenic complications. These patients also have an increased acuity that contributes to the stresses of the ICU environment. This requires quality services from collaborative disciplines in order to decrease mortality and morbidity.

The critical care environment can be overwhelming to the patient who is bombarded by stimuli from all aspects. The equipment is strange and emits lights and sounds. These continuous stimuli subject patients to sensory overload in these areas, yet with decreased stimuli to the other senses, such as smell and touch, leading to sensory deprivation (48). The effects of the constant noise and stimuli from the critical care environment provide little opportunity for sleep, thus increasing confusion and disorientation. Medications can also affect the sensorium, particularly in elderly patients. Metabolism of medications is prolonged in the elderly population and toxic effects can be seen more readily as well as prolongation of the effects.

Sleep deprivation in the critical care patient usually occurs within 2–7 days of admission. One sleep cycle is 60–120 minutes in length and the body usually requires four to five complete sleep cycles at night (49). The 24-hour environment of the ICU, combined with the critical illness of the patient requiring constant monitoring and assessing, leads to interruption of these cycles. Physiological alterations as a result of sleep deprivation include: alteration in temperature, decreased muscle strength, altered chemistries, altered electroencephalogram, decreased sympathetic nervous system function, increased levels of stress hormones and their effects, and decreased levels in energy metabolism. The psychological effects can be: illusions, visual/auditory hallucinations, inability to maintain a line of thought, disorientation, confusion/psychoses, inattention, agitation/combativeness, and noncooperation (50). These effects can prolong recovery and persist after discharge from the ICU.

Sensory overload also has a severe effect on the physiological and psychological well-being of ICU patients. Physiological effects have similar effects of stress such as: dilated pupils, peripheral vasoconstriction, increased blood pressure, increased muscle tension, and increased adrenalin release as a result of stress. Psychological effects can be seen as feelings of helplessness and hopelessness, decreased ability to fall asleep, sleep disturbances, reversal of the sleep cycle from night to day, hallucination, headaches, and emotional lability (51).

The decibel level of sounds in the ICU can be misleading because of the overwhelming number of sounds present. The decibel level of rush hour traffic is 90 decibels (dB), whereas opening a package of sterile gloves is 86 dB, and a patient coughing is 80 dB. Generally, sounds that are higher pitched, intermittent, of long duration and impulse in character are more annoying—e.g., cardiac monitor alarm, ventilator alarm, etc. (51). One study of hospital environmental noise found that the noisiest places in the hospital were patient care environments such as the recovery room, the critical care units, and incubators (52).

Nurses can have a tremendous impact on reducing noise. Measures such as installing insulation, carpets, and draperies, maintaining separate patient rooms, utilizing separate nursing stations from patient areas, centralized monitors, keeping equipment away from the bedside or head of bed, restricting communication at the bedside to that with the patient, and monitoring noise level and movement of personnel in the unit can decrease the effects of sensory overload (51).

The sleep cycle of patients should be monitored and facilitated. Such small interventions as dimming lights, closing doors and curtains, allowing naps during the day, grouping care during the night to give 1–2 hours of uninterrupted sleep, and providing adequate sedation will decrease morbidity and reduce length of stay when allowing adequate rest and healing time (51).

The therapeutic aspect of touch cannot be overlooked. In the high technological intensity of the critical care environment, patients can sometimes get lost in the rush to get tasks done. Allowing families to touch and have access to the patient is also important to the psychological well-being of the patient, especially infants. One study evaluated the contact of newborns with caregivers and found almost all contacts were with staff members, were brief but occurred frequently, yet they received infrequent social contact such as holding, rocking, and being talked to. All of these have been found to have particular developmental significance (53). Touch is, thus, another important aspect of the critical care environment.

TECHNOLOGICAL FACTORS

Critical care medicine presents an ever-changing technological environment. The technology with which patients are treated carries with it the risk of iatrogenic complications. These may include complications associated with the insertion or maintenance of hemodynamic monitoring catheters (e.g., pneumothorax, infection, hemmorhage), the attainment and maintenance of airway and adequate oxygenation (e.g., laryngospasm), or the administration of medication and its potential effects on every organ system (54,55). Nosocomial infection is not uncommon in the ICU. Maki (56), in his review article, cited the numerous variables associated with infection (e.g., host susceptibility, invasive/surgical procedures, length of ICU stay, and microbe resistance).

As technology has advanced, the environment within the critical care units has also changed. New machines, tests, and available resources such as organ donation have had an impact on the critical care environment to the point where patients are suffering from "high tech" and the lack of "high touch." These changes affect not only patients, but also their families, the nursing staff taking care of them, and the system of planning for the care of patients.

One critical area in technology advances in critical care is that of organ transplants. Organ donation research started as early as 1905 when Carrel and Guthrie transplanted the heart of a small dog onto the neck of a large dog and it started beating (57). By the end of the 1960s, most major organ systems had been transplanted (58). Subsequently, advances have focused on suppression of rejection and on developing better techniques and equipment for procuring and transporting organs, thereby allowing organs to remain viable for longer periods of time.

With transplantation becoming commonplace, a host of transplant centers for various organs has emerged. This rapid growth has diluted the pool of available donors as well as raised concerns about regulations of transplant centers. The National Organ Transplant Act of 1984 addressed organ procurement agencies, established a national network for disbursement of donated organs, and evaluated reimbursement and medication payment practices (58).

Publicity relating to the need of specific children for organs has heightened public awareness of the need for donations. The number of individuals needing organs far surpasses the number of organs available for donation. However, reports are frequently heard that families are grateful for the opportunity to donate organs in order to achieve some meaning out of a senseless accident (trauma, drowning, etc.). The Required Consent Law and its implementation as reviewed by Medicare has increased the awareness of health care staff members regarding the viewing of terminal patients as possible donors (58). Health care providers are frequently reluctant, however, to approach families in times of crisis and grief to ask for organs, especially multiple organs. The use of organ donation cards are helpful to identify patients who have previously made a deliberate decision to donate organs.

A study evaluating attitudes regarding transplantation in England found that 30% of the public was concerned that physicians would be pressured into removing organs before they were sure the patient was dead (59). Defining death and establishing when a potential donor is dead can be difficult and fraught with legal pitfalls, especially in cases involving the patient as a victim of an act on the part of another. Establishing brain death based on the Harvard criteria is simple and straightforward; however, many potential organ donors do not meet those strict guidelines and other legally acceptable criteria then need to be developed. This may delay transplantation, thus possibly losing organs to infection or ischemia or losing the recipient to organ failure (60,61).

Organ transplantation in neonates has become more feasible, although available organs are extremely rare. Anencephalic infants have been viewed as potential organ donors because their life-span can only be a few weeks. Current law requires that brainstem function must cease and brain death criteria must be met before organs can be used. Thus, many organs will be lost due to hypoxic injury during the brain death process (62). Strict criteria for obtaining organs, for determining which organs are suitable for transplant, and for deciding which patients are candidates for donation have been clearly defined (58, 63, 64).

The issues of reimbursement for cost of transplantation and its follow-up care are at the core of ethical issues involved in transplantation. Frequently, transplant programs will require cash payments before candidates will be accepted into a program. Some transplant costs are reimbursable through insurance, such as kidney transplants. The costs of long-term follow-up, especially the medications, are staggering, and allocation of resources to benefit the most people may dictate that fewer funds will be available for transplants. The State of Oregon, for instance, has recently reallocated monies from the transplant program to prenatal care, reasoning that preventive care is more cost-effective.

CONCLUSION

The critical care environment may be an overwhelming situation for the patient. The patient, however, is the integral focus of the environment because he or she is the purpose for which the environment is created. Reviewing the trends and influences that characterize the critical care patient enables the nurse manager to evaluate appropriateness of services as well as to formalize programs that will foster quality care.

REFERENCES

1. American Heart Association. Textbook of Advanced Cardiac Life Support. Dallas: AHA, 1987.
2. Eisenberg M, Bergner L, Hallstrom A. Paramedic programs and out-of-hospital cardiac arrest: I. Factors associated with successful resuscitation. Am J Public Health 1979;69:30–38. In: Hunt RC, McCabe JB, Hamilton, GC, Krohmer JR. Influence of emergency medical services systems and prehospital defibrillation on survival of sudden cardiac death. Am J Emerg Med 1989;7:68–82.
3. Baxt WG, Moody P. The differential survival of trauma patients. J Trauma 1987;27:602–606.
4. Moylan JA, Fitzpatrick KT, Beyer AJ, Georgiade GS. Factors improving survival in multisystem trauma patients. Ann Surg 1988;207:679–685.
5. Dean NC, Haug PJ, Hawker PJ. Effect of mobile paramedic units on outcome in patients with myocardial infarction. Ann Emerg Med 1988;17(10):1034–1041.
6. Naughton MJ, Brissie RM, Bessey PQ, McEachern MM. Demography of penetrating cardiac trauma. Ann Surg 1989;209(6):676–681.
7. Wright SW, Dronen SC, Combs TJ, Storer D. Aeromedical transport of patients with post-traumatic cardiac arrest. Ann Emerg Med 1989;18(7):721–726.
8. Urdaneta LF, Miller BK, Ringenberg BJ, Cram AE, Scott DH. Role of an emergency helicopter transport service in rural trauma. Arch Surg 1987;122(9):992–996.
9. Burney RE, Rhee KJ, Cornell RG, Bowman M, Storer D, Moylan J. Evaluation of hospital-based aeromedical transport programs using therapeutic intervention scoring. Aviat Space Environ Med 1988;59(6):563–566.
10. Girotti MJ, Pagliarello G, Todd TR, Demajo W, Cain J, Walker P, Patterson A. Physician-accompanied transport of surgical intensive care patients. Can J Anaesth 1988;35(3):303–308.
11. Rettig PC, Markus GR, Bentley J, et al. Medicare's prospective payment system: the expectations and the realities. Inquiry 1987;24:173–188.
12. Ahmad M, Fergus L, Stothard P, Harrington D, Sivak E, Farmer R. Impact of diagnosis-related groups' prospective payment of utilization of medical intensive care. Chest 1988;93:176–179.
13. Sloan FA, Morrisey MA, Valvona J. Medicare prospective payment and the use of medical technologies in hospitals. Med Care 1988;25:837–853.
14. Mushlin AI, Panzer RJ, Black ER, Greenland P, Regenstreif DI. Quality of care during a community-wide experiment in prospective payment to hospitals. Med Care 1988;26:1081–1091.
15. DesHarnais SI, Chesney JD, Fleming ST. Should DRG assignment be based on age? Med Care 1988;26:124–131.
16. Calore KA, Iezzoni L. Disease staging and PMCs: Can they improve DRGs? Med Care 1987;25:724–737.
17. Munoz E, Rosner F, Friedman R, Sterman H, Goldstein J, Wise L. Financial risk, hospital cost, and complications and comorbidities in medical non-complications and comorbidity stratified diagnosis related groups. Am J Med 1988;84:933–939.
18. Sorrentino EA. Hospitals vary by LOS, charges, reimbursements and death rates. Nurs Man 1989;20:54–60.
19. Freund D, Shachtman RH, Ruffin M, Quade D. Analysis of length-of-stay differences between investor-owned and voluntary hospitals. Inquiry 1985;22:33–44.
20. Munoz E, Barrios E, Johnson H, et al. Pediatric patients, race, and DRG prospective hospital payment. AJDC 1989;143:612–616.
21. Corcoran L, Diers D. Nursing intensity in cardiac surgical care. Nurs Man 1989;20:80I–80P.
22. Trofino J. JCAHO nursing standards nursing care hours and LOS per DRG Part I. Nurs Man 1989;20:29–32.
23. McCloskey JC. Implications of costing out nursing

services for reimbursement. Nurs Man 1989;20:44–49.
24. Akin BV, Rucker L, Hubbel FA, Cygan RW, Waitzkin H. Access to medical care in a medically indigent population. J Gen Intern Med 1989;4(3):216–220.
25. Hamnor ML, Lalor LJ. The aged patient in the critical care setting. Focus Crit Car 1983;10(6):23–29.
26. Fedullo AJ, Swinburne AJ. Relationship of patient age to cost and survival in a medical ICU. Crit Car Med 1983;11(3):155–159.
27. Strauss MJ, LoGerfo JP, Yeltatzie JA, Temkin N, Hudson LD. Rationing of intensive care unit services: an every day occurrence. JAMA 1986;255(9):1143–1146.
28. Study findings show that elderly patients respondto critical care medicine. Critical Care Forum: Highlights from the 18th Annual Meeting of the Society of Critical Care Medicine: 1,4.
29. Van der Merwe WM, Collins JF. Acute renal failure in a critical care unit. 1989;102(863):96–93.
30. Crump JM, Duncan DA, Wears R. Analysis of multiple organ system failure in trauma and nontrauma patients. Am Surg 1988;54(12):702–708.
31. Raffin TA. Intensive care unit survival or patients with systemic illness. Am Rev Respir Dis 1989;140:S28–35.
32. Fry DE. Multiple system organ failure. Surg Clin North Am 1988;68(1):107–122.
33. Darling FE, Duff JH, Mustard RA, Finley RJ. Multiorgan failure in critically ill patients. Can J Surg 1988;31(3):172–176.
34. Wilkinson JD, Pollack MM, Glass NL, Kanter RK, Katz RW, Steinhart CM. Mortality associated with multiple organ system failure and sepsis in pediatric intensive care unit. J Pediatr 1987;111(3):324–328.
35. Torrecilla C, Cortes JL, Chamorro C, Rubio JJ, Galdos P. Dominguez de Villota E. Prognostic assessment of the acute complications of bone marrow transplantation requiring intensive therapy. Intens Care Med 1988;14(4):393–398.
36. Gibney RT, Stollery DE, Lefebvre RE, Sharun CJ, Chan P. Continuous arteriovenous hemodialysis: an alternative therapy for acture renal failure associated with critical illness. Can Med Assoc J 1988;139(9):861–866.
37. Weiss L, Danielson BF, Wikstrom B, Hedstrand U, Wahlberg J. Continuous arteriovenous hemofiltration in the treatment of 100 critically ill patients with acute renal failure: report on clinical outcome and nutritional aspects. Clin Nephrol 1989;31(4):184–189.
38. Brazilay E, Kessler D, Berlot G, Gullo A, Geber D, Ben Zeev I. Use of extracorporeal supportive techniques as additional treatment for septic-induced multiple organ failure patients. Crit Care Med 1989;17(7):634–637.
39. Bihari DJ, Tinker J. The therapeutic value of vasodilator prostaglandins in multiple organ failure associated with sepsis. Intens Car Med 1988;15(1):2–7.
40. Sheagren JN. Mechanism-oriented therapy for multiple system organ failure. Crit Care Clin 1989; 5(2):393–409.
41. Baur M, Schmid TO, Laudauer B. Role of phospholipase A in multiorgan failure with special reference to ARDS and acute renal failure (ARF). Klin Wochenschrift 1989;67(3):196–202.
42. Gattinoni L, Solca M, Pesenti A, Marcolin R, Riboni A, Gavazzeni B, Bassi F, Guiffrida A, Prato P. Combined use of artifical lung and kidney in the treatment of terminal acute respiratory distress syndrome. Life Support Systems 1983;1:365–367.
43. Blaufuss J, Wallace J. Proceedings from the 16th Annual National Teaching Institute, American Association of Critical Care Nurses, May 15-19, 1989;372–373.
44. Shortridge L. Patterns of morbidity and mortality during pregnancy and infancy. In: Valanis B (ed). Epidemiology in Nursing and Health Care. Connecticut: Appleton Century Crofts, 1986:p 139–165.
45. Resnick MB, Carter RL, Ariet M. Effect of birth weight, race, and sex on survival of low-birth-weight infants in neonatal intensive care. Am J Obstet Gynecol 1989;161:184–187.
46. Saule H, Riegel K, Beltinger C. Effectiveness of neonatal transport systems. J Perinat Med 1987;15:515–521.
47. McCormick MC. Long-term follow-up of infants discharged from neonatal intensive care units. JAMA 1989;261:1767–1772.
48. Hansell HN. The behavioral effect of noise on man: The patient with "intensive care unit psychosis." Heart Lung 1984;13(1)59–65.
49. Fernsebner B. Sleep deprivation in patients. AORN 1983;37(1):35–42.
50. Snyder-Halpern R. The effect of critical care unit noise on patient sleep cycles. CCQ 1985;149(8):41–50.
51. Baker CF. Sensory overload and noise in the ICU: sources of environmental stress. CCQ 1984;149(8):66–79.
52. Falk S, Woods NF. Hospital noise-levels and potential health hazards. N Engl J Med 1973;289:774–481.
53. Gottfied AW, Hodgman JE, Brown KW. How intensive is newborn intensive care. An environmental analysis. Pediatrics 1984;74(2):292–294.
54. Khan FA. Common complications in critically ill patients. Disease-a-Month 1988;34:221–293.
55. Sladen A. Complications of invasive hemodynamic monitoring in the intensive care unit. Curr Probl Surg 1988;25:69–145.
56. Maki DG. Risk factors for nosocomial infection in intensive care [Commentary]. Arch Intern Med 1989;149:30–35.
57. Harbison SP. Origins of vascular surgery: The Carrel-Guthrie letters. Surgery 1962;52:406.
58. Felks-McVay R. Transplantation. In: Kinney MR, Packa DR, Dunbar SB (eds). AACN's Clinical Reference for Critical-Care Nursing, 2nd ed. New York: McGraw-Hill Book, 1988: p 1556–1596.

59. Wakeford RE, Stepney R. Obstacles to organ donation. Br J Surg 1989;76(5):436–539.
60. Peabody JL, Emery JR, Ashwal S. Experience with anencephalic infants as propective organ donors. N Engl J Med 1989;321(6):344–350.
61. Soifer B, Gelb AW. The multiple organ donor: identification and management. Ann Intern Med 1989;110(10):814–823.
62. Kozlowski LM. Case study in identification and maintenance of an organ donor. Heart Lung 1988;17(4):366–371.
63. Goodman MR, Aung MH. Cerebral death: Theological, judicial, and medical aspects. Heart Lung 1978;1:477–483.
64. Rudy E. Brain death. Dimens Crit Care Nurs Nurs 1982;1(3):178–184.

Appendix A

STRUCTURE STANDARDS

The American Association of Critical-Care Nurses (AACN) in *Standards of Nursing Care of the Critically Ill* has promulgated the following Structure Standards in relation to the provision of supplies and equipment to the critically ill patient:

IV. Comprehensive Standard: The critical care nurse shall have essential equipment, services, and supplies immediately available at all times.

    IV.a. Supporting Standard: The critical care nurse shall participate in establishing an inventory of necessary equipment and supplies for each unit that will:
- Include routine as well as emergency equipment;
- Reflect the specific needs of the potential patient population;
- Be reviewed annually.

    IV.b. Supporting Standard: The critical care nurse shall participate in establishing written policies and procedures for ordering, monitoring, and replacing equipment, medication, and supplies needed for each unit.

    IV.c. Supporting Standard: The critical care nurse shall ensure that equipment, medications, and supplies considered necessary during emergency situations shall:
- Be centrally located and readily accessible;
- Have documented inspection once each shift and after each use by appropriate personnel.

    IV.d. Supporting Standard: The critical care nurse shall be responsible for ensuring the availability of necessary supplies and equipment before admission of a new patient.

    IV.e. Supporting Standard: Provision shall be made for replenishment of needed supplies on a 24-hour basis.

    IV.f. Supporting Standard: The critical care nurse shall have knowledge of and access to available clinical and laboratory services that may be necessary during emergency situations.

    IV.g. Supporting Standard: The critical care nurse shall demonstrate knowledge of and reponsibility for obtaining necessary equipment, medications, and supplies.

Appendix B

AACN POSITION STATEMENT

American Association of Critical-Care Nurses □ One Civic Plaza, Newport Beach, CA 92660 □ (714) 644-9310

Collaborative Practice Model: The Organization of Human Resources in Critical Care Units

COLLABORATION HAS BEEN IDENTIFIED as a pivotal component in the delivery of quality health care. The Joint Commission on Accreditation of Hospitals acknowledges the importance of collaboration in critical care units by requiring that the activities of such units be guided by a multidisciplinary approach that includes both nursing and medical input.[1] The National Commission on Nursing also urges collaboration by proposing as an immediate goal that trustees and health care administrators "promote and support complementary practice between nurses and physicians" and that they "examine organizational structure to ensure that nurse administrators are part of the policy-making bodies of the institution and have authority to collaborate on an equal footing with the medical leaders in the institution."[2] This should clearly extend to the unit level where care is delivered.

Recognizing that the impetus for true collaboration must originate with health care professionals themselves, the Board of Directors of the American Association of Critical-Care Nurses and the Council of the Society of Critical Care Medicine commissioned a task force of experienced critical care practitioners, managers, and educators to identify principles by which critical care units could successfully function through collaboration. Because different unit environments and hospital structures exist, no model is likely to be universally applicable. However, regardless of these differences, certain principles must be incorporated into the unit structure in order to assure optimal functioning of the unit.

Principles

1. Responsibility and accountability for effective functioning of a critical care unit must be vested in physician and nurse directors who are on an equal decision-making level.
2. These directors must be appropriately prepared and educated. In addition to competence in patient management, they need knowledge and experience in the following areas: management principles, resources management, and skills in interpersonal relationships (including conflict resolution).
3. The organizational structure of a critical care unit must ensure that physicians are autonomous when dealing with issues that affect medical practice.
4. The organizational structure of a critical care unit must ensure that nurses are autonomous when dealing with issues that affect nursing practice.
5. Some aspects of patient care require interdependence between physicians and nurses. These aspects must be identified and addressed jointly.
6. Every critically ill person requires medical and nursing care. The services of additional disciplines may also be required in specific situations. In order to provide a holistic approach, the care delivered by other health team members must be coordinated by the physician and nurse directors.
7. Unit support services must be organized to enable the directors to optimally carry out their primary responsibilities in the practice of their respective disciplines (i.e., patient care).

8. The directors are accountable for the evaluation of the quality and efficiency of care and the financial provision of that care. They must develop a unit-specific system for the evaluation of care on a timely basis.
9. The directors are responsible for creating and maintaining an environment in which individuals have opportunities to realize their potentials.
10. Close collaboration between the directors is essential for successful management. This collaboration can be enhanced by daily rounds, weekly meetings, and other means that will ensure continuous, open communication.

REFERENCES

1. Joint Commission on Accreditation of Hospitals: Accreditation manual for hospitals. Chicago, 1982, The Commission, p. 182.
2. National Commission on Nursing: Initial report and preliminary recommendations. Chicago, 1981, The Commission, p. 62.

Adopted by AACN Board of Directors and SCCM Council, October 1982

Appendix C

AN EXAMPLE OF A VISITORS' INFORMATION GUIDE*

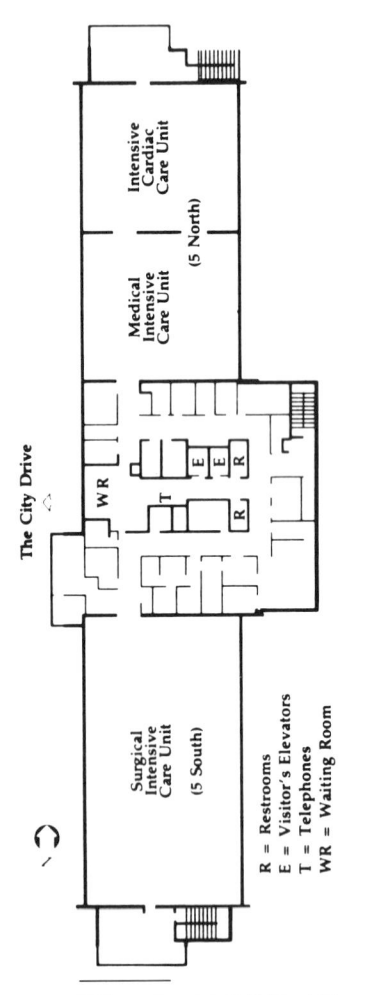

Dignity for our patients. Compassion for their families and loved ones. An enduring dedication to provide the finest critical care possible. These are the hallmarks of UCI Medical Center's Adult Critical Care Units.

The Units provide:

- 32 private patient rooms.
- Open, centralized nursing stations, providing a view of the patients at all times.
- Glass windows in all rooms—to permit observation while maintaining the privacy of each patient.
- Sophisticated computerized monitoring and support equipment.
- Non-threatening, pleasant atmosphere.
- Quiet rooms and private consultation rooms for families.

Your Telephone Directory of Services and Support

Surgical Intensive Care Unit (SICU)	634-5311
Medical Intensive Care Unit (MICU)	634-5308
Intensive Cardiac Care Unit (ICCU)	634-5325
Social Services (Lodging, transportation and other community services)	634-5644
Business Office	634-6355
Private Patient Coordinator	634-5678
Chaplain Services	634-6112

Art reproduced by permission of the Foundation for Critical Care

UCI MEDICAL CENTER
101 The City Drive
Orange, California 92668

UCI MEDICAL CENTER

ADULT CRITICAL CARE VISITOR INFORMATION

"In critical care, it strikes me that the issues are three: realism, dignity and love."

SENATOR JACOB K. JAVITS

R = Restrooms
E = Visitor's Elevators
T = Telephones
WR = Waiting Room

Surgical Intensive Care Unit (5 South)
Medical Intensive Care Unit
Intensive Cardiac Care Unit (5 North)
The City Drive

---

* From the Adult Critical Care Nursing Service Task Force 1988. Adult Critical Care Visitor Information. University of California, Irvine Medical Center, Orange, CA.

For Your Peace of Mind

We invite you to take a closer look at the UCI Medical Center Adult Critical Care Units.

Let this brochure be your guide. If you have any questions, please ask the nurses or doctors who are caring for your loved one.

All Around You

When you visit the Adult Critical Care Units, you're surrounded by the latest medical technology. Some of the machines and equipment you may see during your visit might seem impersonal. But remember, the medical equipment is there to support and sustain the patients or to permit the skilled physicians and nurses of the units to constantly monitor each patient's condition.

Inside Information

Within the Adult Critical Care Units are:

1. Surgical Intensive Care (SICU)
2. Medical Intensive Care (MICU)
3. Intensive Cardiac Care (ICCU)

Complementing these units are the adjacent Cardiac Catheterization Lab and Respiratory Therapy Department. For guests and visitors, we've provided two "quiet rooms" where you may wait in comfort—as well as consultation rooms where you can talk privately with doctors, social workers or other support staff who are here to help families and patients.

We Understand

We know that this can be a very difficult time for you. Stress and anxiety can take their toll. So we urge you to take care of yourself—and to take heart in knowing that the person you care about is receiving the finest care possible.

Realism, Dignity, Love

The health of our patients and your own comfort and peace of mind are vitally important to us. Our constant goal is to keep you informed, honestly and realistically, of the patient's condition and outlook.

You might want to write down the names of those medical professionals who are closest to your loved one. These are the people who can help you and keep you advised.

Attending Physician: _____

Resident Physician: _____

Primary Nurse: _____

Nurse Manager: _____

We encourage you to designate one individual as family spokesperson. The staff will keep the spokesperson informed of the patient's condition.

Important Information for You

To ensure dignity for the patient and to help you through this difficult period, we ask for your cooperation.

Visitation

All visitors must check in at the entrance to the 5th floor by calling the unit's nursing station. Two visitors maximum at one time, please. We ask for your understanding that a patient's condition, physician's rounds or other situations may cause us to restrict visits.

Children

Children under 16 are not allowed on the 5th floor.

Coordination

You will find it helpful to check with the nursing staff to help coordinate your visits.

Telephone

Limited information will be provided over the phone. If the patient or spokesperson requests privacy, all information can be withheld.

Cleanliness

Please wash your hands in the sinks provided before entering and leaving the patient's room.

Personal Property

The family should take the patient's personal belongings home.

Appendix D

INTERVENTIONS AND STRATEGIES RELATED TO CARING FOR THE FAMILY OF THE CRITICALLY ILL PATIENT*

COGNITIVE NEEDS
1. *To know specific facts about the patient's progress*
   - Avoid using generalizations such as "he's much better" or "things are about the same today."
   - Use the same simple terms each day to discuss progress and concerns, e.g., heart rhythm, blood pressure, level of pain, level of oxygen in the blood, not responding/very groggy/sleepy/awake, etc. This permits families to focus their attention on the same frame of reference and, when applicable, to put closure on the topic.
   - Relate progress to the illness as you have described it initially.
   - Use Kardex care plan to communicate phrases and areas of concern being discussed so that all staff members use the same terminology.
2. *To know the probable outcome*
   - Be as realistic as possible, but be aware of family's coping mechanisms (such as denial) as well as their need for hope.
   - Establish short-term goals so that positive change can be identified.
   - If patient's prognosis is poor, allow adequate time to spend with the family so that feelings or questions can emerge. Establish times to meet with the family again.
   - Verbally recognize and accept family's desire for certainty in an ambiguous situation.
3. *To know exactly what is being done for patient (how the patient is being treated medically, why things are being done for the patient)*
   - Briefly describe each line and/or monitoring device including i.v.s, urinary catheters, arterial lines, NG tubes, O₂ devices, etc.
   - Encourage questions.
   - Remember that one explanation may not be enough; high anxiety is a barrier to learning.
   - Use simple terminology such as "breathing tube," "cardiogram," or "special intravenous" rather than ET tube, ECG, or Swan.
   - Base explanations of treatment on patient's illness as you have described it initially. Reinforce explanations of pathophysiology as needed.
   - Promote continuity through Kardex care plan.
4. *To have questions answered honestly*
   - Be specific; discuss all issues as they relate to the patient as a unique individual.
   - Maintain good communications with MDs so that you are aware of what they have told the family. Discuss with them any information they feel should be withheld to determine rationale.
   - Use Kardex for consistency. Families quickly become aware of evasive answers. Consistency decreases anxiety and feelings of dehumanization and promotes cooperation.
5. *To have explanations given in understandable terms*
   - Assess family's knowledge base and previous experience with patient's condition.
   - Determine family's priority for learning; what do they want to know *immediately*.
   - Provide basic pathophysiology slowly, allowing time for questions. Repeat pathophysiological concepts when discussing treatments or progress.
   - Divide informational sessions so that family is not overwhelmed. Establish times at which you will meet with them again to answer questions or provide additional explanations.
   - Remember that high anxiety is a barrier to learning; explanations usually have to be repeated.
   - Respond positively to questions, recognizing their right to understand what is happening.
6. *To have explanations about the environment before going into ICU for the first time*
   - Utilize time before patient arrives on unit to meet with family.
   - Remember that the environment *is* frightening, not only equipment surrounding patient but also the equipment used for other patients in the area.
   - While describing equipment, reinforce that much of it is present for prevention and early detection of problems.
   - Verbally recognize family's feelings and reassure them that these are normal and acceptable.
   - Be with family during their first visit to provide additional explanations and support.

---

* From Bouman CC. Identifying priority concerns of families of ICU patients. Dimens Crit Care Nurse 1984;3(5):316–317.

- For planned admissions (e.g., postsurgery): Establish guidelines in teaching for discussion of environment;
  Provide patient and family tours of ICU if desired;
  Have ICU nurse who will be assigned to patient immediately postoperatively meet with family before surgery.

EMOTIONAL NEEDS

1. *To be assured that the best possible care is being given to the patient*
   - Emphasize that nurses and MDs are experienced and have special training to provide expert care.
   - Verbally recognize family's anxiety. Reassure them that such feelings are normal and that they do not have to hide these feelings.
2. *To be called at home about changes in the patient's condition*
   - Establish guidelines with family about who is to be called and under what timeframe.
   - Always ensure that the family knows about any significant change in patient's status before first visit of the day. If they are not informed about a change for the better, families frequently assume they will not be notified of other changes.
3. *To receive information about the patient once each day*
   - Set aside one period during the day, as early as possible, during which progress reports are given.
   - Assess whether this time is adequate and follow-up as needed particularly if patient's condition changes.
   - Be aware if family needs to have MD's explanations clarified or amplified.
4. *To see that patient frequently*
   - Review policy for visiting hours to determine if they meet family needs; adjust as permitted by current situation in ICU.
   - If a visit must be delayed, explain reason clearly; ensure that family does not think it is due to a crisis unnecessarily.
   - Provide time for patient and family to be alone together.
5. *To feel that hospital personnel care about the patient*
   - Utilize information gained during family assessment to anticipate concerns.
   - Listen to their concerns, demonstrating that staff sees patient as an individual. Seek advice from family about meeting patient needs.
   - Tell family what patient communicates.
   - Focus on sensations patient will feel and possibly communicate to family so that these do not come as a surprise (e.g., sore throat after ET tube, fragility of chest wall after CABG, noisy environment, etc).
   - Remember that family members may focus on small details such as the patient's position, cleanliness, or bedding because these are within their normal range of control.
   - Remember that family members may take out their feelings of grief, guilt, or anger on nursing staff; be patient.
6. *To talk about the possibility of patient's death*
   To feel there is hope
   - Determine family's perception of situation.
   - Anticipate and recognize signs of anticipatory grieving. Provide open-ended questions to encourage them to express their feelings.
   - Provide concrete information about patient's status slowly, assessing family's response. False reassurance and, the opposite extreme, abrupt confrontation with reality may hinder appropriate coping.
   - Meet jointly with MD and family to discuss use/nonuse of extraordinary measures or advanced life support systems.
   - Be aware of how comfortable you are when discussing a patient's death. If unable to meet family needs by yourself, arrange for another member of the health team to meet with family.
   - If there is no *real* contraindication, allow family to spend additional time with dying patient.

PHYSICAL NEEDS

1. *To help with the patient's physical care*
   - Anticipate that family may be reluctant to touch patient because of equipment and yet want/need to touch him/her for reassurance. Small directed activities meet both these concerns.
   - Allow family to do small things for patient, e.g., providing ice chips, wiping face, washing old blood off hands, etc.
   - If there is no *real* contraindication, allow family member to provide parts of a.m. care.
2. *To visit anytime*
   - Discuss visiting regulations with family
   - Be flexible if special arrangements need to be made because of other responsibilities.
   - Recognize that this need will probably diminish over a few days as families become aware that visiting hours are adequate and that they *will* be permitted to see patient frequently.
3. *To have the waiting room near ICU*
   To have a telephone near the waiting room
   - If these are not available in your hospital, be aware of potential stress for families. Discuss possible alternative arrangements.
   - Discuss the need for family members to get adequate rest so that they will be able to support patient later in hospitalization and to minimize extended stays by family in waiting room.
4. *To have a place to be alone in the hospital*
   - Direct family to hospital chapel if available.
   - Determine if there is space available nearby, currently being used for another purpose, which could be used by family for periods of privacy.
   - If weather is pleasant, encourage family to exercise outside during the day.

Index

Page numbers in *italics* denote figures; those followed by "t" denote tables.

A

Accountability, 36
Acquired immune deficiency syndrome, 42–43, 122
  diagnosis of, 43
  economic impact of, 15
  incubation period for, 43
  nosocomial infections and, 43
  pathophysiology of, 43
  transmission of, 43
Administrative sphere, 6, 29–60. *See also* specific topics
  clinical nursing research program, 52–60
  definition of, 6
  infection control, 42–51
  quality control, 30–41
Aging, 14, 165–166
Air conditioning, 111
Air transport of patients, 162–163
Ambulatory services. *See* Outpatient services
American Association of Critical-Care Nurses
  "Collaborative Practice Model" of, 148, 174–175
  environmental model of, 3, *3*
Ancillary services, 63–73. *See also* specific topics
  functions of, 63
  nutritional therapy, 69–71, 71t
  occupational therapy, 68–69, 70t
  pharmacy, 63–66
  physical therapy, 68, 69t
  relationships with, 146
  respiratory therapy, 66–68
  social work and discharge planning, 71–73, 72t–73t
Anesthetic gases, 127–128
Antimicrobial agents, resistance to, 44
Antineoplastic agents, 126
Architectural design, 107–108. *See also* Plant design
Argyris's immaturity-maturity continuum, 142
Aspergillus, 44, 46
Average length of stay, 163–165

B

Bacteremias, nosocomial, 47
Bacterial infections, 44–45
  in transplant patient, 48–49
Beneficial occupancy, 109. *See also* Plant design
  correcting punch list items and, 109
  definition of, 118
  equipment inventory for, 109
  interdepartmental impact of, 109
  space orientation before, 109
  training on new equipment and, 109

Biomedical services, 74–75
  consultation guidelines for, 74, 75t
  dependence on, 74–75
  functional areas of, 74

C

Campylobacter, 124
Candida, 44
Cardiopulmonary resuscitation, 162
Cardiovascular prosthesis infection, 48
Case/procedural carts, 92
Central nervous system infections, 48
Chemical dependency, 134–135
Chickenpox, 123–124
Cleaning agents, 127
Clinical nurse specialist, 23, 149–150
  relationships of, 149–150
  research mentoring of staff by, 55
  role of, 149
Clinical nursing research program, 52–60
  administrative readiness for, 54–56
    indicators of, 55, 55t
    mentoring for, 55
  benefits of, 52–53
    positive impact in practice setting, 53
    professional image enhancement, 52–53
    recruitment and retention, 52
    staff development, 53
  establishing proper climate for, 57
  future perspectives on, 59
  Guidelines for the Investigative Functions of Nurses, 53, 54t
  initiation of, 57–59
    focus groups, 58
    journal clubs, 58
    practicality of research question, 59
    replication studies, 59
    selecting participants, 57
    selecting research question, 58, 58t
    time commitment, 57–58
  reasons for failure of, 59
  resources for, 56–57
    emotional, 57
    financial, 56–57
    personnel, 56
Closed circuit television monitoring, 110
Collaboration, 8, 62–63, 147–154
  continuum of unity, 62, *63*
  in critical care, 148–151
    clinical nurse specialist, 149–150
    critical care nurse manager, 150

181

Collaboration—*continued*
   critical care staff nurse, 149
   expanded nursing roles, 151
   nursing case manager, 150
   physicians, 151
   self-destructive relationships, 148
   sources of ideas, 149
   static relationships, 148
   synergistic relationships, 148–149
   technical/support and assistive personnel, 150–151
  hierarchy of elements in, 62, *63*
  managed care model and, 62–63
  models of, 147, *147–148*
  nurse manager's promotion of, 151–153
   interactions with indirect care givers, 153
   interactions with other direct care givers, 153
   interactions with physicians, 153
   intraorganization, 153–154
   intraunit, 152–153
   leadership and management style, 152
   team building, 152, 152t
  synergy and, 147
  teamwork and, 147–148
  in work groups, 144–146
"Collaborative Practice Model," 148, 174–175
Colonization, 42, 121
Communication, 79
  critical care unit systems for, 111
  interactional, 87
  materiel support systems and, 86–87
  matrix, *86*, 86–87
  with patient's family, 157–159
  transactional, 87
Computers, 93–103
  definition of, 94
  functions in critical care unit, 93–94
  future of, 102
  hardware for, 94, *95*, 95t
  HELP system, 97, *98*, 102
  history of hospital use of, 93
  interface with users, *96*
  for nursing information system, 96–101. *See also* Nursing information system
  for nursing management system, 101–102. *See also* Nursing management system
  for quality assurance program, 102
  security in, 102, 103t
  sharing resources of, 95–96
  software for, 94
  types of, 95
  usefulness of, 103
   to critical care units, 103
   to nurse managers, 103
   to organizations, 103
   to staff nurses, 103
Construction, 108–109. *See also* Plant design
Consultation
  with biomedical services, 74, 75t
  with nutritional support service, 70–71
  with occupational therapist, 69, 70t
  with pharmacist, 64, 65t
  with physical therapist, 68, 69t
  with respiratory therapist, 66, 68t
Consultation rooms, 114
Contamination rooms, 114
Contractors, 108. *See also* Plant design
Cost containment, 35
Critical care environment
  characteristics of, 36
  conceptual model of, 2–8
   of AACN, *3*, 23
   administrative program sphere of, 6
   critically ill patient in, 8
   differentiating spheres of, 4–5, *5*
   human resource sphere of, 7–8
   impact of previous models on, 2–3
   physical sphere of, 7
   purpose of, 2
   strategic sphere of, 5–6
   support systems sphere of, 6–7
  definition of, 2
  elements in, 2
   interrelatedness of, 4
  family's relationship to, 157, *158*
  health hazards in, 120
  impact on patient, 8, 167–169
  literature and research on, 3–4, 6
Critical Care Nursing Standards, 33, 33t, 173
Critical care patient, 162–170
  definition of, 8
  effects of critical care environment on, 8, 167–169
   noise, 168
   psychological impact, 168
   sensory overload, 168
   sleep deprivation, 168
   touch, 168–169
  factors affecting survival of, 162–163
  population demographics and, 165–167
   chronic health problems, 166
   increasing life expectancy, 165–166
   survival of premature infants, 167
  prehospital CPR for, 162
  socioeconomic factors affecting, 163–165
   Diagnosis Related Groups, 163–165
   medically indigent patients, 165
   reduction in average length of stay, 163–165
  technological factors affecting, 169–170
   iatrogenic complications, 169
   organ transplants, 169–170
  transport of, 162–163, 167
Critical care unit
  costs of, 34
  functions of computers in, 93–94, 103
  guide for visitors to, 176–177
  mortality rate in, 42
  nosocomial infections in, 42–51. *See also* Infection control; Nosocomial infections
  physical design of, 49, 106–114. *See also* Plant design
  reasons for admission to, 34
  relationships in, 146. *See also* Relationships
Cryptococcus neoformans, 44
Cryptosporidium enteritis, 44

Culture, 19
  definition of, 19
  values and, 19
Customer relations, 8, 156–160
  focus of, 157
  identifying family needs, 156–157, *157–158*
  nurse manager's role in, 160
  patient/family feedback and evaluation, 160
  principles of, 157–160
    human interaction, 157–158
    physical and policy manipulation, 158–159
    resource development, 159–160
Cytomegalovirus, 44
  as occupational hazard, 124
  during pregnancy, 124
  transmission of, 124
Cytotoxic agents, 126

D

Debridement rooms, 114
Diagnosis Related Groups, 12, 163–165
  evaluation of, 164
  nursing effects of, 165
  research on, 60
Diagnostic services, 63, 77–78
Discharge planning, 72–73, 73t
Drug hazards, 126

E

Elderly patients, 14, 16, 165–166
Electrical current, 129
  amperes of, 130
  conductors of, 129–130
  definition of, 129
  sensitivity to, 130–131
Electrical power
  design of, 111
  education about, 130
  hazards of, 129–131
Emergency Medical System, 162
Endocarditis, postoperative, 48
Endoscopic rooms, 114
Enterobacter, 44
Environmental services, 75–76
  functional areas in, 75t
  hallmarks of good service, 76t
  infection control and, 75–76
Epstein-Barr virus, 44
Equipment
  basic list of, 112
  designing layouts for, 112
  on headwalls, 112–113
  impact on critical care patient, 168
  inventory of, 109
  methods of purchase of, 114
  in nursing station, 113
  protective, 121
  providing access to, 112
  training on, 109
Equipment procurement, 114–118
  by competitive proposal process, 117
  direct control for, 117
  by direct purchase, 116
  by donations, 117–118
  by group purchase agreement, 116–117
  request for proposal process for, 114–116
    elements of, 114–115
    flexibility of, 114
    hospital's terms and conditions, 115
    quotation, 116
    scope of work, 115
    statement of purpose, 115
    supporting documentation, 116
    technical specifications, 115–116
  by reuse of existing equipment, 118
  timing of, 118
Escherichia coli, 44
Exchange carts, 91–92
Exercise rooms, 113

F

Facilities management, 73–74
  definition of, 73
  functional areas of, 73, 74t
  standards for, 73–74
Family, 156–160
  identifying needs of, 156–157, *157*
  interactions with, 157–159
  relationship to critical care environment, 157, *158*
  resource development in support of, 159–160
  soliciting feedback from, 160
  strategies for caring for, 178–179
    cognitive needs, 178–179
    emotional needs, 179
    physical needs, 179
  visitors' information guide for, 176–177
  wayfinding activities of, 158–159
Federal Hazard Communication Standard, 125
Fee-for-service payment model, 5
Foot-candles, 110
  definition of, 118
Fungal infections, 44

G

Gases, 126–128
  compressed, 126–127
  waste anesthetic, 127–128
Goals and objectives, 25–26, *26–27*
Grief rooms, 114
Group dynamics, 144–146
"Guidelines for Establishing Joint or Collaborative Practice in Hospitals," 148
Guidelines for the Investigative Functions of Nurses, 53, 54t

H

Hand-washing, 123
Headwalls, 108, 110
Health care reform, 15–16, 163–165
Health insurance, 12–14
  impact on patient care, 14
  lack of, 14, 165
  Medicare, 12–14
  private, 14

INDEX

Health Organization-Environment Model, 2–3
Heating ventilation, 111
HELP computer system, 97, 98, 102
Hemodialysis, continuous arteriovenous, 167
Hemofiltraion, continuous arteriovenous, 167
Hepatitis
 blood transmission of, 46
 as occupational hazard
  hepatitis A, 125
  hepatitis B, 122–123
  non-A, non-B hepatitis, 123
 in transplant patient, 49
 vaccination against, 122
Herpes simplex, 44
 as occupational hazard, 123
 reactivation of, 123
 transmission of, 123
 in transplant patient, 49
Herpes zoster, 44
 as occupational hazard, 123–124
 in transplant patient, 49
Herpetic whitlow, 123
Herzberg's satisfiers-dissatisfiers, 142
Histoplasma capsulatum, 44
Human immunodeficiency virus, 43
 diagnosis of, 43
 as occupational hazard, 122–123
 transmission of, 43, 46
Human needs theories, 140–142
 Argyris' immaturity-maturity continuum, 142
 Herzberg's satisfiers-dissatisfiers, 142
 Maslow's hierarchy, 141, *141*
 McGregor's theory X and theory Y, 142
Human resources sphere, 7–8, 139–170. *See also* specific topics
 critical care patient, 162–170
 customer relations, 156–160
 definition of, 7
 roles and relationships, 140–154
Hydrotherapy rooms, 114

I

Immunodeficiency
 acquired diseases of, 42. *See also* Acquired immune deficiency syndrome
 definition of, 42
 inherited diseases of, 42
Infection control, 49–51. *See also* Nosocomial infections
 design of critical care unit and, 49
  single patient cubicles, 49
  traffic control, 49
  ventilation system, 49
 nurse manager's role in, 50
 program for, 49–50
  components of, 50
  goal of, 49
  priority for, 49–50
 quality assurance activities for, 50
 role of environmental services in, 75–76
 surveillance and reporting for, 50–51
Information systems, 76–77. *See also* Nursing information system

computers for, 93
functions of, 76–77
 administrative system, 78t
 clinical system, 77t
security in, 102, 103t
steering committee for implementation of, 96–98
types of, 76
usefulness in critical care unit, 76–77
Interior design, 107. *See also* Plant design
Interpersonal relationships. *See* Relationships; Role
Intestinal pathogens, 124–125
Isapora belli, 44

J

Job description, 143
Junction box, definition of, 118

K

Klebsiella pneumoniae, 44

L

Labor market, 15
Laboratory services, 77–78, 79t
Laminar airflow, 111
Lasers, 133
 classification of, 133
 definition of, 133
 eye exposure to, 133
 hazards of, 133
Laundry service, 76
Legionella pneumophila, 46
Liability. *See* Risk management
Lifting, as occupational hazard, 129
Lighting
 design of, 110
 as occupational hazard, 129
Listeria, 45
 in transplant patient, 49

M

Magnet Hospital Study, 4, 6–7
Malnutrition. *See also* Nutritional support services
 complications of, 69
 risk factors for, 71t
Managed care, 62–63, 79–80
 benefits of, 80
 client-centered conferences for, 80
 focus of, 80
 goals of, 62
 plan for, 79
Maslow's hierarchy of needs, 141, *141*
Material Safety Data Sheets, 125–126
Materiel support systems, 7, 82–92
 analysis, design and implementation of, 82–83, *83*
 communication and, *86*, 86–87
 establishing goals and objectives of, 83
 establishing standards and specifications for, 84–86
 nurse manager's goal relating to, 82
 participants in designing of, 83–84
 planning for quality system, 82
 policies and procedures of, 86
 provision alternatives in, 91–92
  case/procedural carts, 92

exchange carts, 91–92
  par level transfer system, 91
  requisition system, 91
 quality control in, 88–91
  absolutes of quality, 88
  document for, 89, *90*
  incident/problem resolution techniques, 89–91
  performance criteria, 88
  performance monitoring system, 88–89
  responsibility for system integrity, 88
  statistical process control, 89
 shared educational experiences and, 87–88
McGregor's theory X and theory Y, 142
Medicaid, 13, 165
Medicare, 12–14
 costs for, 13–14
 payments by, 165
 percentage of revenue from, 13
 profit margins of, 13
 Prospective Payment System of, 12–14, 163–165
Meningitis, 48, 124
Microorganisms, 43–45
 antibiotic resistance of, 44
 dose of, 44
 pathogenicity of, 44
 portal of exit for, 45
 reservoir for, 45
 source of, 45
 types of, 44–45
 virulence of, 44
Mission and philosophy statement, 20–21, 21t
Multiple system organ failure, 165–167
Mycobacterium, 45, 125

N

National Commission on Nursing, 4
National Electrical Code, 131
National Organ Transplant Act, 169
Neisseria meningitidis, 124
Nocardia asteroides, 45
Noise, 131
 control of, 131
 definition of, 131
 hazards of, 131
 in ICU, 168
Nominal group technique, 20
Norms, 19–20
 definition of, 19
 values and, 19–20
Nosocomial infections, 42–51, 169
 chain of, 42
 common types of, 46
  bacteremias, 47
  implantable prosthetic devices, 47–48
  pneumonia, 46–47
  urinary tract infection, 47
 control of, 49–51. *See also* Infection control
 cost of, 42
 definition of, 42
 factors required for, 42
 host factors in, 42–43
  immunodeficiency states, 42–43

 nutritional status, 43
 severity of illness, 43
 microbial agent factors in, 43–45
  antibiotic resistance, 44
  dose, 44
  genetic make-up, 44
  pathogenicity, 44
  portal of exit, 45
  reservoir, 45
  source, 45
  types of agents, 44–45
  virulence, 44
 mode of transmission of, 45–46, 121
  airborne, 46
  blood, drugs, food, water, 46
  direct contact, 45
  droplet spread, 45–46
  indirect contact, 45
  vectors, 46
 prevalence of, 42
 risk factors for, 42
 in transplant patient, 48–49
 vs. colonization, 42, 121
 vs. community-acquired infections, 42
Nurse manager, 150
 AACN position statement on, 7–8
 customer relations function of, 160
 environmental linkages of, 2
 infection control function of, 50
 leadership and management style of, 152
 quality control function of, 30–41
 responsibilities of, 7
 roles of, 7, 150–153
  conceptual model of, 2
  strategic planning by, 18–27. *See also* Strategic planning
  as team builder, 150–153, 152t
  use of computers by, 103
  use of conceptual model by, 2–8
Nursing administration research, 6
Nursing case manager, 150
Nursing information system, 96–101. *See also* Information systems
 definition of, 97
 factors causing delay in development of, 96
 implementation of, 99–100
  conversion methods, 99–100
  planning for, 99
  timetable for, 101t
 integrated vs. stand-alone approaches to, 97, *98*
 management of changes due to, 99
 postimplementation evaluation of, 101
 steering committee for implementation of, 96–98
  membership of, 97–98
  objectives of, 98
  responsibilities of, 98
 training program for, 100
 user groups for, 98–99
Nursing management system, 101–102
 budget, 102
 components of, 101
 nursing staffing and scheduling systems, 101–102
Nursing station design, 113

INDEX

Nutritional status, infection and, 43
Nutritional support services, 69–71
  assessment for, 69
  consultation guidelines for, 70–71
  metabolic support goals, 69, 72t
  nutritional risk factors, 69, 71t
  standards for, 69–70

O

Objectives, 25–26, *26–27*
Occupational hazards, 120–135
  behavior-related, 121
  biological, 121–125. *See also* Nosocomial infections
    hepatitis A, 125
    hepatitis B, 122–123
    herpes viruses, 123–124
    human immunodeficiency virus, 122–123
    intestinal pathogens, 124–125
    meningococcus, 124
    modes of transmission of, 121
    non-A, non-B hepatitis, 123
    pregnant nurses and, 121
    preventing transmission of, 121–122. *See also* Infection control
    scabies, 125
    susceptibility to, 121
    tuberculosis, 125
    universal precautions for, 122–123
  chemical, 125–128
    antineoplastic agents, 126
    compressed gases, 126–127
    exposure to, 126
    Material Safety Data Sheets about, 125–126
    other drugs, 126
    soaps and cleaning agents, 127
    solvents, 127
    waste anesthetic gases, 127–128
  classification of, 120
  engineering controls for, 120–121
  ergonomic, 128–129
    lifting, 128
    lighting, 129
    work station hazards, 129
  identification of, 120
  job training and, 120
  personal protective equipment for, 121
  physical, 129–133
    electrical, 129–131
    lasers, 133
    noise, 131
    radiation, 131–132
  psychological, 133–135
    chemical dependency, 134–135
    shift work, 134
    stress, 133–134
    violence, 135
  substitution for, 120
Occupational therapy, 68–69
  consultation guidelines for, 69, 70t
  services of, 68–69, 70t
Organizational behavior, 140. *See also* Relationships; Role

Outpatient services, 12–14
  factors encouraging, 16
  impact of technology on, 14
  price controls on, 12–13
Oxygen cylinders, 127

P

Patient care incident, 36
  causes of, 36
  definition of, 36
  indicators related to, 37, 37t
  investigation of, 36
  reporting of, 36–37
Patient care space, 110
Payment models, 5, 12–14
Pharmacy services, 63–66
  clinical role of pharmacist, 63–64
  consultation guidelines for, 65t
  distribution system, 64–66
    intravenous admixture method, 64–65
    unit-dose method, 64–65
  drug cost control, 65–66
  operational scope of, 65t
  purpose of, 63
  responsibility for therapeutic drug monitoring, 66
  unit-based pharmacist, 65–66
Physical sphere, 7, 105–135. *See also* specific topics
  definition of, 7
  equipment procurement, 114–118
  occupational hazards, 120–135
  plant design, 106–114
Physical therapy, 68
  consultation guidelines for, 68, 69t
  services of, 68, 69t
Physician relationships, 146–147, 151, 153
Planning. *See* Strategic planning
Plant design, 106–114
  developmental process for, 107–109
    architectural design and creation of construction documents, 107–108
    beneficial occupancy and move-in logistics, 109
    conception of need, 107
    construction, 108–109
    definition of design, 107
    making project work, 109
    selection of contractors and suppliers, 108
  future development and, 114
  glossary for, 118–119
  infection control and, 49
  key factors in, 109–114
    communications, 111
    electrical power, 111
    equipment, 112–113
    heating ventilation and air conditioning, 111
    lighting, 110
    nursing station, 113
    patient care space, 110
    plumbing, 111–112
    public space, 114
    support space, 114
    treatment rooms, 113–114
  mock-up of, 106–107
  nurses' role in, 106

planning team for, 106–107
    goals of, 106
    membership of, 106
    subcommittees of, 106
Plumbing, 111–112
Pneumocystis carinii, 44
Pneumonia, nosocomial, 46–47
    causes of, 46–47
    factors predisposing to, 47
    incidence of, 46
    prevention of, 47
Population demographics, 165–167
    aging, 14, 165–166
    chronic health problems, 166–167
    growth, 14
    increasing life expectancy, 165–166
    survival of premature infants, 167
Preferred Provider Organizations (PPOs), 14
Premature infants, survival of, 167
Prospective Payment Assessment Commission, 13
Prospective Payment System, 5, 12–14
    economic impact of, 13, 163–165
    effect on quality of life and outcome, 164
    establishment of, 12
    exemption from, 12
    payments of, 12
Prosthetic devices, infections in, 47–48
Protozoal infections, 44
Pseudomonas aeruginosa, 44
Punch list, 118–119
    compilation of, 109, 118
    definition of, 119

Q

Quality assurance, 30–33
    computerization of program for, 102
    definition of, 30–31
    evolution of, 30
    goals of, 30–31
    incident/problem resolution techniques, 89–91
    indicators for, 33t
        thresholds for, 32–33
    for infection control, 50
    inservice training on, 33
    methodology of, 32
    monitoring for, 32
        methods of, 32
        purpose of, 32
        scheduling of, 32
        statistical process control, 89
        tool for, 32–33
    nurse manager's role in, 31, 33
    participation in, 31–32
    plan for, 31–32
    responsibility for, 30
    sources for evaluation, 32
    standards for, 32
Quality control, 6, 30–41, 88–91
    absolutes of, 88
    activities included in, 30. *See also* specific topics
        integration of, 30
        processes of, 30
        quality assurance, 30–33

risk management, 30, 35–37
utilization control, 30, 33–35
definition of, 30
determining responsibility in, 88
establishing performance criteria for, 88
establishing performance monitoring system for, 88–89
indicators of, 30
nurse manager's role in, 30, 37, 40
relationship to environment, 30, *31*
at unit level, 37–41, *40*
    benefits of, 41
    medical-nursing collaboration for, 39–40
    report form for, *38–39*
    staff involvement in, 37–39
value and, 16
Quality control committee, 32, 38–39
Quality surveys, 88–89

R

Radiation hazards, 131–132
    distance and, 132
    duration of exposure and, 132, 132–133
    effects of shielding, 132
    physical effects of exposure, 132
    of portable x-ray machines, 132
    radiation exposure, 131–132
Radiology, 77–78, 79t
Rehabilitation services, 68–69
Relationships, organizational, 8, 145–154. *See also* Role
    collaborative, 147–154. *See also* Collaboration
    external, 146–147
        interdepartmental, 146
        with physicians, 146–147
    impact on unit work and patient outcome, 154
    internal, 146
        interunit, 146
        intraunit, 146
    in work groups, 144–146, *145*
Research
    establishing clinical nursing research program, 52–60. *See also* Clinical nursing research program
    historical, 3–4
    on nursing administration, 6
Respiratory therapy, 66–68
    clinical role of respiratory therapist, 66
    consultation guidelines for, 66, 68t
    operational scope of, 67t
    standards for, 67–68
Retrospective payment model, 5
Risk, definition of, 35
Risk management, 6, 30, 35–37
    characteristics of critical care environment and, 36
    definition of, 35
    indicators for, 36
    methods of, 36
    nurse manager's role in, 36
    patient care incidents and, 36–37, 37t
    quality assurance audits and, 37
Role. *See also* Relationships
    of clinical nurse specialist, 149
    collaborative, 148–151. *See also* Collaboration

Role—*continued*
  concept of, 142–143
  expanding nursing roles, 151
  of nurse manager, 2, 7, 150–153
  of nursing case manager, 150
  perception of, 143–144
    group, 143
    individual, 143–144
    organizational, 143
  role conflict, 144, 144t
  role set, 144
  of staff nurse, 143, 149
  task behavior and, 143
  vs. position, 143
  work groups and, 144–145, *145*
Rotavirus, 124

S

Safety practices. *See also* Occupational hazards
  patient care outcomes and, 7
  standards of, 7
  training in, 120–121
Salmonella, 124
Scabies, 125
Sensory overload, 168
Serratia, 44
Shift work, 134
Shigella, 124
Shingles, 123–124
Sleep deprivation, 168
Soaps, 127
Social Security Amendment of 1972, 30
Social work, 71–73
  consultation guidelines for, 72t
  discharge planning and, 72–73, 73t
  responsibilities of, 72
  services of, 72t
Society of Critical Care Medicine, 6
Solvents, 127
Staff development, 53
Staff nurse, 149
  definition of, 149
  relationships of, 149
  role of, 143, 149
Staff nurses
  usefulness of computers to, 103
Standards, 32
  examples of, 33t
  for facilities management, 73–74
  for materiel support systems, 84–86
  for nutritional support services, 69–70
  outcome, 32
  process, 32
  for respiratory therapy, 67–68
  structure, 32, 173
Staphylococci, 44–45
Strategic planning, 18–27
  critical success factors in, 22
  definition of, 5, 21–22
  elements in, 19, *19*
  evaluation of, 26
  formulating strategies in, 23–25, *24*, 25t

  goals and objectives of, 25–26, *26–27*
  identifying strengths and weaknesses in, 22–23, 23t
  planning for, 18–19
  professional value identification in, 19–21, 21t
  responsibility for, 18
  scanning external environment in, 21–22
  structure of document for, 27
Strategic sphere, 5–6, 11–17. *See also* specific topics
  definition of, 5
  elements of, 12–16
    aging of population, 14
    impact of AIDS, 15
    labor market, 15
    Medicare and other payment models, 12–14
    new health care system, 15–16
    other external forces, 16
    quality, 16
    technology, 14–15
  impact on critical care, 16–17
  strategic planning and evaluation of, 21–22
Strategies
  definition of, 23
  duration of, 19
  examples of, 24–25, 25t
  formulation of, 23–25, *24*
  related to caring for patient's family, 178–179
  work sheet for, 26, *27*
Strengths and weaknesses, 22–23, 23t
Stress, 133–134
  control of, 134
  definition of, 133
  health effects of, 134
Study on the Efficacy of Nosocomial Infection Control, 42
Styles Stipulation, 62
Substance abuse, 134–135
Supply carts, 91–92
Support services. *See also* specific topics
  biomedical services, 74–75, 75t
  collaboration with, 62–63, *63*
  environmental services and laundry, 75t–76t, 75–76
  facilities management, 73–74, 74t
  functions of, 63
  information systems, 76–77, 77t–78t
  relationships with, 146
  unintegrated vs. integrated health care and, 63, *64*
Support systems sphere, 6–7, 61–103. *See also* specific topics
  ancillary and support services, 62–80
  computer technology, 93–103
  definition of, 6
  materiel services, 82–92
  necessity for, 6
Surgical rooms, 114
System, 82–83
  definition of, 82
  life cycle of, 82, *83*
    analysis and design phase, 83
    conceptual phase, 82–83
    implementation phase, 83
    operational and control phase, 83
    subsystems of, 82

T

Tax Equity and Fiscal Responsibility Act of 1982, 5
Team building, 150–153, 152t
Teamwork, 147–148. *See also* Collaboration
Technical/support personnel, 150–151
Technology, 14–15
  computers, 93–103. *See also* Computers
  managerial considerations related to, 7
  nursing research on, 53
Third party payors, 14
Touch, 168–169
Toxoplasma gondii, 44
Transplantation, 169–170
  attitudes about, 169
  history of, 169
  infections and, 48–49
  multiple system organ failure and, 166
  in neonates, 169–170
  network for donated organs for, 169
  payment for, 170
  persons in need of, 169
  requesting donations for, 169
Transport of patients, 162–163
  by air, 162–163
  maternal-neonatal, 167
  survival and, 163
Treatment room design, 113–114
Tuberculosis, as occupational hazard, 125

U

Universal precautions, 122–123
Urinary tract infections, nosocomial, 47
  catheterization and, 47
  complications of, 47
Utilization control, 30, 33–35
  costs of critical care units, 34
  definition of, 34
  documenting benefits of critical care, 34
  evaluation methods for, 35
  impetus for, 33–34
  indicators for, 34–35, 35t
  objective of, 34
  reasons for admission to critical care unit, 34
  review activities of, 34

V

Values, 19–21
  acquiring of, 19
  culture and, 19
  nominal group technique for identification of, 20
  norms and, 19–20
  strategic planning and, 19–21
Varicella zoster virus, 123–124
Violence, 135
  control of, 135
  definition of, 135
  education about, 135
  health effects of, 135
Viral infections, 44
Visitors' information guide, 176–177

W

Waiting rooms
  design of, 114
  family needs related to, 156–159
  physical amenities in, 159
Wayfinding, 158–159
Weaknesses and strengths, 22–23, 23t
Work groups, 144–146, *145*
Work station hazards, 129